WORLDSCAPES

SECOND EDITION

Allan C. Ayers, Douglas King, Mary F. Pirie,
Ian Watson, Ken Whitcombe

General Editor: T.H. Masterton

Oliver & Boyd

Acknowledgements

The publishers wish to thank the following for permission to reproduce photographs:

Daily Telegraph Colour Library: European Space Agency/Meteosat (1, p6); ZEFA (12, p11; 14, p12; 34, p55; 28, p120); J Topham (15, p13; 29, p121 Christine Osborne); Kenbarry Productions (17, 18, p14; 18, p26; 27, p29; 30, p30; 38, p35; 2, p78; 15, p116); Frank Spooner (23, p17 Gamma); Colorific (24, p17 Cindy Karp/Visions; 33, p54 Jim Balog/Black Star; 29, p92 Brian Boyd; 3, p101 David Burnett/Tigray; 34, p123 Urve Kuusik); Eric Kay (9, p21; 11, p22; 12, p23; 15, p25; 26, p29; 21, p50); A C Waltham (10, p21; 13, p23; 25, p29; 31, p31; 33, p32; 22, p70; 27, p120); Frank Lane Agency (17, p26 M Nimmo); Aerofilms (22, p28; 30, p121); Bryan and Cherry Alexander (35, 36, p34); Barnaby's Picture Library (39, p35; 26, p51); Tony and Marion Morrison (40, p36; 41, p59; 44, 45, p60); J Allan Cash (43, p37; 32, p54; 24, 25, p71; 28, 29, p72; 26, p91; 30, p93; 38, p98; 7, p114); Western Americana (7, p41); Mansell Collection (10, p42; 13, p45; 17, 18, p46; 20, 21, p70; 1, p79); Bruce Coleman (21, p49 Jonathan Wright); Douglas Dickins (26, p51); Joyce Mackenzie (27, p51); South American Pictures/Bill Leimbach (36, p56; 37, 38, p57; 43, p60); Snowy Mountain Scheme (7, 8, p64); Australian Information Service, London (9, p65; 13, p66); D A Sherriff (11, 12, p66); G R Roberts (15, p67); Landform Slides (19, p67); Hong Kong Government (23, p71; 30, p73; 33, p74; 34, 35, p75; 36, 37, p76; 38, p77); Geoslides (27, p72; 8, p114); Ursula Whitely (32, p74); The Photo Source (12, p84); Irish Tourist Board (16, p86); International Society for Educational Information Tokyo, Inc (22, 23, p89; 24, 25, p90; 27, p91); J Hillelson Agency (32, p94 Dr Georg Gerster); Abu Dhabi Petroleum Company (8, 9, p104); Massey Ferguson (11, 12, p107); Joanne Thoms, Eccles Primary School, Duns (2, p112); Camera Press (5, p112 Jan O Tornquist; 17, p116 T A Wilkie; 24, p119); FAO (6, p113); R R Furness (9, p114); International Whaling Commission/Durant Hembree (10, p114); National Coal Board (11, 12, p115); North of England Open Air Museum (14, p115); British Waterways Board (18, p116); Wade Cooper Associates (19, 20, p117); Allan Ayers (21, 22, p118; 26, p120); Everest Double Glazing (23, p118); James Barr (25, p120); Tyne and Wear Passenger Transport Executive (31, p122); Scottish Arts Council/painting by John Byrne (32, p123).

Cover photograph by Image Bank

Illustrated by Barry Adamson, Tim Smith and Cauldron Design Studio

Oliver & Boyd
Robert Stevenson House
1–3 Baxter's Place
Leith Walk
Edinburgh EH1 3BB

A division of Longman Group UK Ltd

First published 1978
Second edition 1987
Second impression 1988
© The General Editor and the contributors 1978, 1987

All rights reserved; no part of this
publication may be reproduced, stored in a retrieval system,
or transmitted in any form or by any means, electronic,
mechanical, photocopying, recording or otherwise, without
either the prior written permission of the publishers
or a licence permitting restricted copying the
United Kingdom issued by the Copyright Licensing Agency Ltd,
33–34 Alfred Place, London WC1E 7DP.

Set in 11/13 Optima Roman and Bold

Produced by Longman Group (FE) Ltd
Printed in Hong Kong

ISBN 0 05 004028 6

Contents

Preface to the second edition   5

1 Earth forces   6
    Zones of the earth   6
    The earth's crust   7
    Parting plates   11
    The value of volcanoes   14
    Sliding plates   15
    Colliding plates   16

2 Nature shapes the landscape   19
    Weathering   19
    Erosion and deposition by running water   21
    Erosion by the sea   25
    Deposition by the sea   27
    The work of moving ice   29
    The work of the wind   32
    Landscapes shaped by climate   33

3 People begin to change the landscape   38
    Introduction   38
    NORTH AMERICA   40
    The first inhabitants   40
    People on the move   42
    Quiz game   44
    End of an Indian way of life   46
    The Plains area today   48
    Surviving Indian lifestyles: the Hopi   50
    A fast-changing mountain landscape: Appalachia   52
    Fast-changing cities   54
    AMAZONIA: DISCOVERY AND DESTRUCTION   56
    The traditional Indian way of life   57
    More recent settlers in Amazonia   58
    What hope for these Indians?   59

4 Landscapes dramatically changed by people   61
    THE SNOWY MOUNTAIN SCHEME   61
    Hydro-electric power   63
    Irrigation   65
    Water for towns and industry   67
    Summarising the Snowy Mountain scheme   68
    HONG KONG   69
    The growth of Hong Kong   69
    Hong Kong today   70
    Housing the population   72
    Trade and industry   74

5 People: patterns and problems   78
    BRITAIN'S POPULATION IN THE PAST AND PRESENT   78
    Birth rate and death rate   79
    The population begins to grow   80
    Understanding population pyramids   82
    Providing for Britain's old people   83
    IRELAND: FALLING POPULATION, 1840–1970   84
    People in Ireland today   86
    JAPAN: A CROWDED INDUSTRIAL COUNTRY   87
    Supporting the population   87
    Using every inch of land   88
    Coastal industry   89
    A high standard of living   90
    Problems with success?   91
    WEST AFRICA: A GROWING POPULATION   92
    Life in a savanna compound   94
    The effect of weather and landscape on farming   96
    Population change in Nigeria   98

6 Aid and trade   99
    Introduction   99
    AID   99
    Emergency aid   99
    Long-term aid   100
    Where does aid come from?   102
    TRADE   103
    Trading in oil   103
    The oil game   106
    Fair trade for rich and poor   107
    World patterns of trade   108
    Rich and poor   109
    Summary   110

7 Using and misusing the environment   111
    What is the environment?   111
    People and the environment   113
    Using and misusing water and air   118
    Ideal environments   120
    Town planning   121
    The street environment   123
    A rural environment under threat   123
    Plan your future   125

Notes for the teacher   127

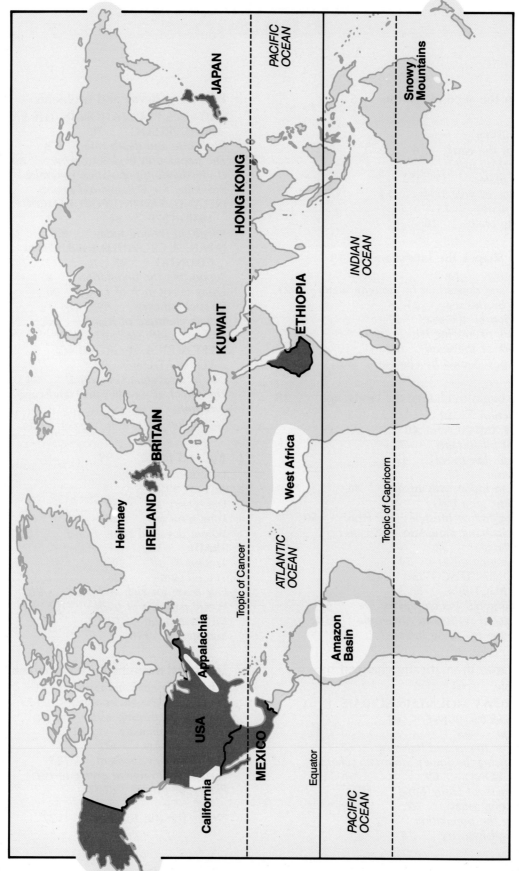

Preface to the second edition

At the time of its publication in 1978, *Worldscapes* was primarily intended to support the then new and radical proposals for the 'common core' geography curriculum in Scottish secondary schools. *Worldscapes*, and its companion volume for the first year, *Homescapes*, together provided a rounded course for Scottish secondary schools which was soon widely adopted.

Shortly after publication it became clear that *Worldscapes* was serving a much wider market than that of the Scottish secondary school, and it was partly due to this wider demand that *Our Landscapes* was published. This offered a substitute for *Homescapes*, for use in schools outside Scotland where the demand was for a broader geographical spread of material than was offered in *Homescapes*. (The latter is now out of print.)

The purpose of this second edition of *Worldscapes* is threefold:

- to update the factual content;
- to redesign and improve the layout in response to comments received from teachers;
- to amend, re-order and add material which offers a fresh selection of studies for use in the modern school curriculum.

In this second edition, Chapter 1 (Earth forces) has been only slightly amended in content since the first edition, and so continues to provide an introduction to earth forces at local and larger scales. The emphasis is on the physical processes and their effects on human activity and welfare.

Chapter 2 (Nature shapes the landscape) has been considerably tightened up and illustrations improved. It continues to provide an introduction to those natural forces which have worked to produce the landscape.

The emphasis in Chapter 3 (People begin to change the landscape) has been redirected. The theme 'developing nature's empty spaces' brings together the concepts of the natural regions with illustrations of continental development.

Chapter 4 (Landscapes dramatically changed by people) strongly emphasises the notion of human impact on the environment. Case studies which appeared in earlier editions have been selectively used and updated.

Chapters 5 and 6 have been greatly revised. The theme of Chapter 5 – an emphasis on population and living standards, illustrated by case studies of Britain, West Africa and Japan – leads sensibly to a discussion of aid and trade in Chapter 6: the distribution of wealth, world problems and resources, and a case study of Sub-Saharan Africa.

Chapter 7 (The environment) has been reduced in length, but the content remains largely unchanged, and so continues to focus on environmental issues of concern to modern society.

1 Earth forces

Zones of the earth

Many years ago, people thought that the earth was flat. It is said that sailors were often afraid to sail too far into unknown seas in case they sailed off the edge of the world. Although people have known for several centuries that the earth is round, it was not until photographs were taken from space that we had visible proof (see photograph 1).

The earth is not a perfect sphere; it bulges slightly at the equator. On a small photograph, however, the bulge does not show. The most you can see is the outline of the earth and the large land masses and oceans.

Although photograph 1 might suggest that the earth is a spherical lump of solid rock, we now know that this is not the case. (Earthquakes and volcanoes which often disturb the earth's surface give us clues to this.) In fact the earth is made up of four different zones as shown in diagram 2.

It is possible to study the earth's **crust** (the outer zone) by drilling, but only the areas near to the surface can be examined in this way. Most of our knowledge of the earth's interior is based on studies of the earth's gravity and magnetic field, and on the ways in which earthquake waves are 'bent' as they pass through the different zones of the earth.

1 The earth from space

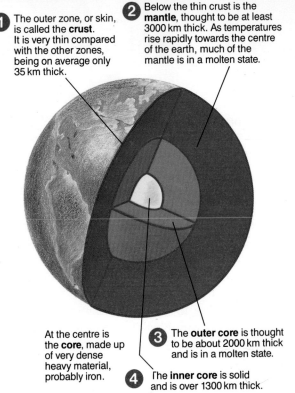

❶ The outer zone, or skin, is called the **crust**. It is very thin compared with the other zones, being on average only 35 km thick.

❷ Below the thin crust is the **mantle**, thought to be at least 3000 km thick. As temperatures rise rapidly towards the centre of the earth, much of the mantle is in a molten state.

At the centre is the **core**, made up of very dense heavy material, probably iron.

❸ The **outer core** is thought to be about 2000 km thick and is in a molten state.

❹ The **inner core** is solid and is over 1300 km thick.

2 The zones of the earth

Core work

1. Use your atlas to help you with these questions.
   (a) What is the name of the land mass in the centre of photograph 1?
   (b) Name the ocean (i) to the right, (ii) to the left of this land mass.
   (c) Name the land mass on the left edge of photograph 1.
2. How can we tell that the earth is not solid?
3. (a) Starting from the outside, name the zones which make up the earth.
   (b) Name the zone which is (i) thickest, (ii) thinnest.
   (c) Which of the zones are (i) molten, (ii) solid?

Extension work

4. Why do you think that it is difficult to study the earth's interior?
5. Name three ways in which scientists can study the earth's interior.

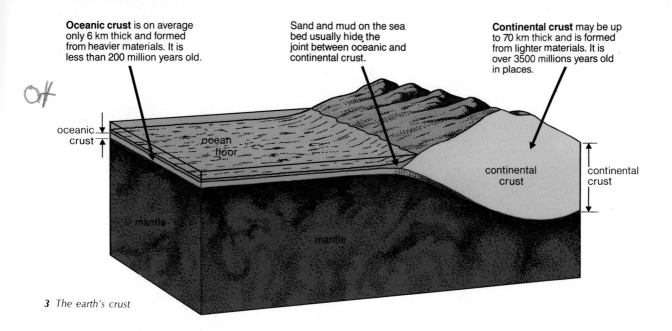

3 The earth's crust

The earth's crust

The earth's crust has been formed by the gradual cooling of the outer layer of the earth since it was formed at least 4500 million years ago. You can perhaps appreciate how long ago this was if you think that human beings have only been on the earth for about four million years.

Recent discoveries have shown that the crust is made up of two different layers, as shown in diagram 3. These discoveries were really the beginning of a new idea, or **theory**, to explain how the crust was formed and how it behaves. This theory is called **plate tectonics**.

The basic idea is that the crust is not one continuous layer. It is made up of separate pieces or **plates** of crust which fit together. These plates vary in size and shape and they seem to be moving slowly about the earth's surface. Some plates are mostly under the sea, and are called **oceanic plates**. The others which are mainly above sea level are called **continental plates**.

Usually plate movements are so slow that they can only be measured by sensitive scientific instruments. Sometimes, however, the movements take place much more quickly and violently, causing **earthquakes**.

Plate tectonics theory helps to explain how, about 250 million years ago, the land masses or continents of the world were all joined together to form one huge land mass called PANGAEA (map 4). This can be proved by looking at evidence of rocks and fossils. Continents which are now separate once fitted together like pieces of a jigsaw (see diagram 5). It seems that the plates on which these continents lie have slowly drifted apart to take up their present positions. This movement is still taking place, and is called **continental drift**.

4 Pangaea

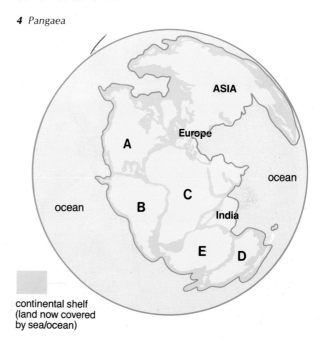

5 Evidence that separate continents were once joined together

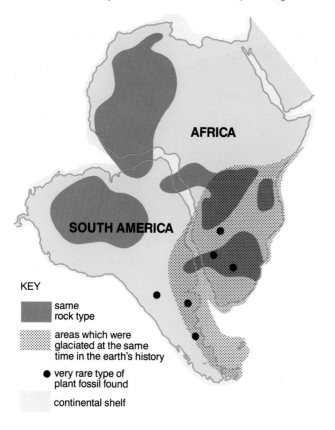

Plate tectonics also help to explain how some of the highest mountains in the world, **fold mountains**, are formed. If you look at maps 6 and 7, you will see that fold mountains are formed near the edges of plates, or **plate boundaries**. Here the crust has been folded or buckled upwards by tremendous pressures. Other mountains rise steeply from the ocean floors, with only their tops rising above the surface of the sea as islands. One of these submarine mountain ranges is shown on map 6, running down the centre of the Atlantic Ocean. It is called the **Mid-Atlantic Ridge** and is formed in a different way from fold mountains.

If you compare maps 6, 7 and 8, you will see that the areas where earthquakes and volcanoes usually occur, and the main mountain areas, are all found within the same broad zones. These zones of movement and buckling mark the edges of the plates which make up the earth's crust. Earthquakes, volcanoes and mountains result from movements along the edges of these plates.

There are thought to be several different ways in which the crustal plates move. Some of them are shown simply in diagram 9.

6 Fold mountain areas of the world, and the Mid-Atlantic Ridge

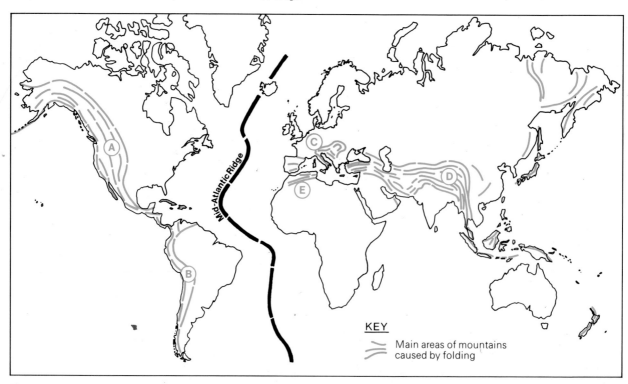

7 The plates of the earth's crust, and their directions of movement

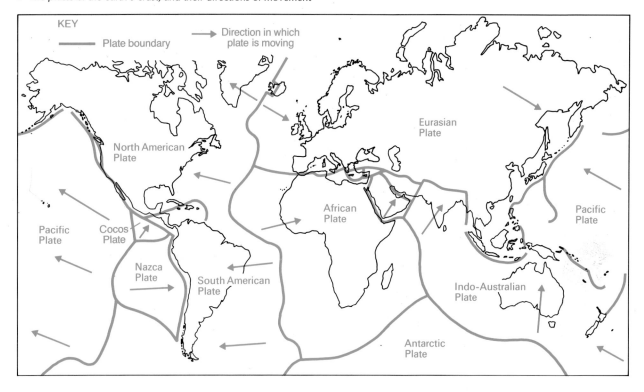

8 Belts of earthquakes and volcanoes

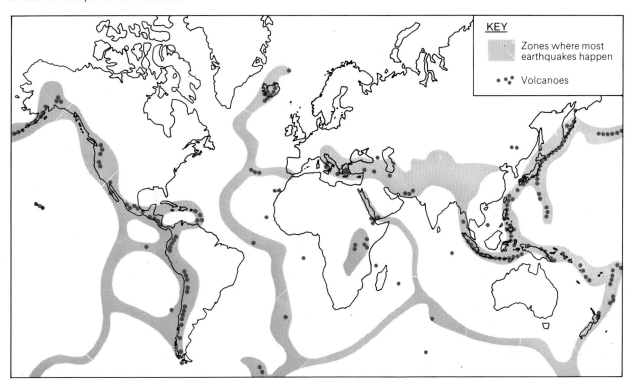

9 *The ways in which plates move*

Parting plates

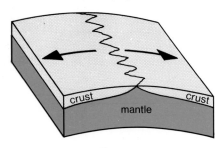

Plates move away from each other

Sliding plates

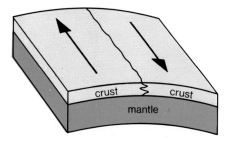

Plates slide past each other

Colliding plates

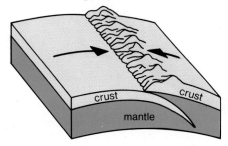

Here, one plate is forced under the other

Core work

6 (a) How old is the earth thought to be?
  (b) How long have people been on the earth?
7 Copy the following table into your workbook. Use diagram 3 to help you complete it.

Type of crust	Oceanic	Continental
Thickness of crust Age of crust Formed of		

8 Use an atlas to find the names of the fold mountains marked A, B, C, D and E on map 6.
9 Compare the position of the fold mountains shown on map 6 with the position of the earthquake and volcano zones shown on map 8. What do you notice?
10 What is the Mid-Atlantic Ridge?
11 Use an atlas and map 8 to name three countries with at least six volcanoes in each.
12 Study map 4 and an atlas. Name areas A, B, C, D, and E that 'broke away' from Pangaea to form five continents.
13 What do the following mean?
Plate tectonics, fold mountains, continental drift, plate boundaries.

Extension work

14 Study diagram 5 carefully. What evidence is there to prove that South America and Africa must have been joined together at one time?
15 Cut up a blank map of the present world. Try to make Pangaea as shown in map 4.
16 Look at map 7 carefully, and the directions in which plates are moving. Over the next several million years what will happen to
  (a) the Atlantic Ocean?
  (b) Australia?
  (c) Iceland?

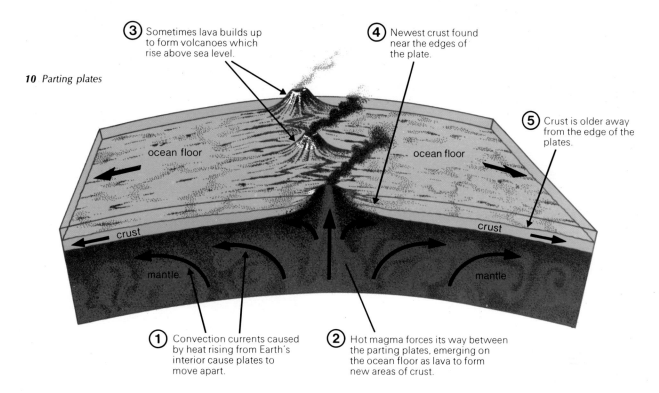

10 Parting plates

③ Sometimes lava builds up to form volcanoes which rise above sea level.

④ Newest crust found near the edges of the plate.

⑤ Crust is older away from the edge of the plates.

① Convection currents caused by heat rising from Earth's interior cause plates to move apart.

② Hot magma forces its way between the parting plates, emerging on the ocean floor as lava to form new areas of crust.

Parting plates

You may remember that the crust under the oceans is usually younger than continental crust. This is because new crust is being formed under the oceans, as shown in diagram 10. When plates are moving apart, new crust forms a ridge along the plate boundary. This is how the Mid-Atlantic Ridge was formed (see maps 6 and 7).

Iceland is an island made up of rock formed by volcanic eruptions (see map 11). It stands on the Mid-Atlantic Ridge between Britain and Greenland. Off the south-west coast of Iceland lies Heimaey, a small island whose inhabitants earn their living mainly by fishing. Most of the inhabitants live in the only town on the island, also called Heimaey.

In the early hours of the morning of 23 January 1973, they suddenly awoke to find that a volcanic eruption was taking place. A giant crack or **fissure** over 400 metres long had opened up in the ground not far from the town, and explosions were blasting red hot lava and black ash up to 100 metres into the air (see photograph 12).

The people quickly packed what they could and hurried to the harbour, where fortunately the fishing fleet was sheltering from a storm.

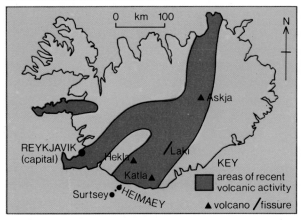

11 Iceland: the main volcanic areas

12 The eruption on Heimaey, January 1973

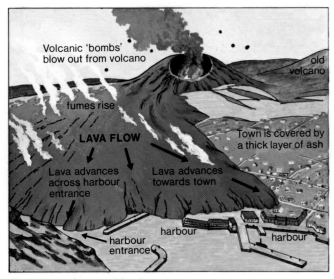

13 Heimaey threatened by the volcano

14 The town of Heimaey shortly after the eruption

Within a few hours they were all safe on the mainland. Only a small force of volunteers stayed behind to save as much property as they could from the eruption.

By the morning, they could see that a new volcano was growing over the crack and that the town was now threatened by four main dangers (see picture 13).

Over the next few weeks, millions of tonnes of lava and ash poured from the volcano. Although the volunteers sometimes had to move out of parts of the town as poisonous gases built up, they worked on. Windows and skylights were covered with corrugated iron to stop volcanic bombs crashing through them. Ash was cleared quickly from each roof to stop the buildings from collapsing under its weight. In an effort to stop the lava from reaching the town and blocking the harbour, bulldozers piled up a huge wall of ash to divert the lava.

At the same time, millions of gallons of seawater were sprayed on the lava front to make it cool and solidify. Although the last major fall of ash was on 16 February, the workers battled against the flow of lava until late in April. Over the next few weeks the volcano became quieter until it was finally declared inactive on 3 July.

As soon as the worst of the eruption was over, work began on clearing the ash away from the streets and from the undamaged buildings. Even before this was finished the islanders were beginning to return (see photograph 15), though this meant living in the shadow of the still-smoking volcano.

The Heimaey eruption was one of the few occasions when people have made a determined attempt to resist a volcanic eruption. Not all of the effects of the eruption were harmful, however. Map 16 shows that the entrance to the harbour is now narrower and more sheltered than it was before the eruption. This is a help to shipping in an area which is often lashed by strong gales. People also stayed on the island throughout the eruption, recording each event as it took place. When information can be collected in this way, it can help scientists to predict future eruptions, perhaps saving many lives.

Core work

17 (a) Why did the people of Heimaey suddenly leave the island in January 1973?
  (b) Study picture 13 carefully. What were the four main dangers facing the town at this time?
  (c) Study photograph 12.
    (i) What two effects of the volcano's eruption can you see in this photograph?
    (ii) How do you think the house became buried?
18 (a) What damage was caused by the eruption?
  (b) Describe the various ways in which the volunteers tried to limit the damage caused by the eruption.
  (c) Describe the scene in photograph 15. What work would the islanders have to do when they returned?
  (d) What good things resulted from the eruption?

15 The islanders return to their homes

Extension work

19 Make up a summary table to show the main stages of the Heimaey eruption and the dates on which they took place. For example, 23 January: a large crack appears in the ground, etc.
20 Draw sketches to show (a) any stage of the eruption; (b) some of the salvage and repair work taking place.

16 Heimaey: the extent of the lava flow

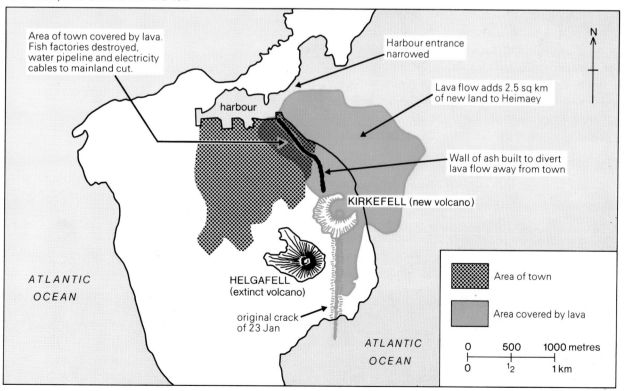

The value of volcanoes

Although many of the effects of volcanic activity are destructive, there can be some useful side-effects. Valuable or useful minerals, such as diamonds or sulphur, are formed by volcanic action. Also, heat from volcanic rocks produces hot water which can be used for heating homes, or for making steam to generate electricity. This is already being done on a small scale in Iceland and New Zealand, but larger projects in many other volcanic areas are being considered.

Some volcanic areas are popular tourists attractions. Many people visit spectacular and beautiful volcanoes such as Vesuvius and Etna in Italy, and Fujiyama in Japan. Hot springs and geysers heated by volcanic action also attract people (see photograph 17). Hot mud is used as a skin beauty treatment.

Solidified lava often breaks down to produce very rich soils which are ideal for farming. Photograph 18 shows such an area in Java, where the fertile soils produce crops year after year. Sometimes, however, these volcanoes erupt again, forcing farmers to flee for their lives. This happened with Mount Etna in 1971.

Rocks formed when molten volcanic material cools and solidifies are called **igneous rocks**. Some of these rocks are very soft: for example, pumice (solidified lava). Others, which are formed underground (for example, granite), are very hard, so they can be used for building when great strength is needed. Granite may also be carved or polished to look decorative.

17 Geyser: Iceland

Core work

21 (a) Name two valuable minerals formed by volcanic action.
   (b) What are hot springs in New Zealand and Iceland used for?
   (c) Why are volcanoes important tourist attractions?

Extension work

22 Some people choose to live near volcanoes, in spite of the danger of further eruptions. Why do you think this is?

23 Try to find out more about some of the famous volcanic areas of the world. Write a paragraph about one or two of these areas. Here are some names to help you: Krakatoa, Surtsey, Stromboli, Mont Pelée, Pompeii, Paricutin, Yellowstone National Park, Mount St Helens.
Find these places in your atlas.

◀ *18* Fertile volcanic soils, Java

Sliding plates

When plates are sliding past each other as shown in diagram 9, the movement is usually so slow that it is hardly noticed. Sometimes, however, the plates stick together for a while and pressures build up. Eventually, the plates free themselves with a jerk which shakes them violently. We call such movements **earthquakes**.

One area where this often happens is along the Pacific coast of the USA, particularly in California. You can find this area in your atlas and on maps 8 and 19.

Most of the movements take place along large cracks in the earth's surface, called **faults**. The largest of these in California is the San Andreas Fault, which can be traced for over 1000 kilometres across the countryside. However, this is only one of the many faults which cross the area. There have been four massive earthquakes since the 1850s and at least one tremor is recorded along the San Andreas Fault every day.

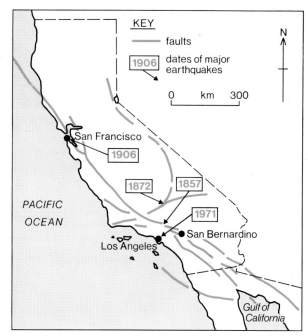

19 California and the San Andreas Fault system

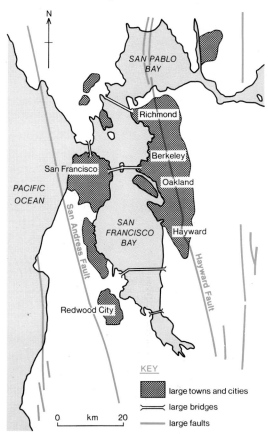

20 San Francisco and major faults

The fact that two of the largest cities in the USA, and many other large towns, are found in this area makes the earthquake problem worse. The 1906 earthquake (and the fires which followed) destroyed much of San Francisco, but at that time the city was only a fraction of the size that it is today. Now, over fifteen million people live in the region affected by the faults (see map 20). Many of them have already experienced small earthquakes and they live in fear of a major disaster.

Scientists know that there will be another major earthquake along one of the faults within the next twenty-five years. But they are not agreed about exactly where and when it will strike.

Core work

24 (a) What is the name of the large fault which cuts through California?
(b) Name the plates which lie on either side of this fault (see map 7).
(c) Why are there so many earthquakes and tremors in this area?
(d) Why might even a small earthquake in this area be a major disaster?

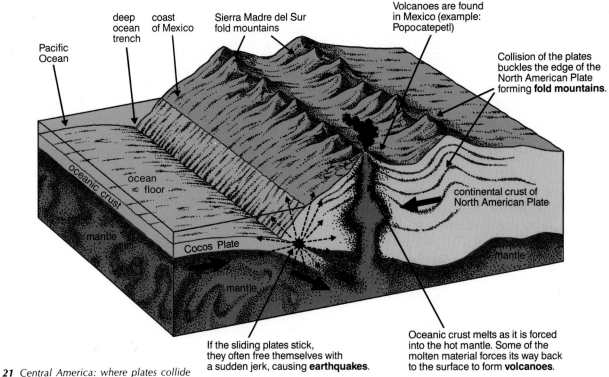

21 Central America: where plates collide

Colliding plates

When plates are moving towards each other and collide, one of the plates is usually forced underneath the other, as shown in diagrams 9 and 21. One area affected by such movements is the western coast of Central and South America. You can find this area in your atlas, and on maps 8 and 22. Normally the movement is very slow, but sudden movements can cause earthquakes.

On 19 September 1985, a massive earthquake shook the whole of south-west Mexico. The **epicentre** of the earthquake was 400 km south-west of Mexico City underneath the Pacific Ocean (see map 22). The earthquake measured 7.8 on the Richter scale, a scale for measuring how strong earthquakes are. The shock waves travelled quickly to Mexico City, the centre of which was built on the soft rock of an old lake and marshes (see diagram inset on map 22). Three minutes later, the ground there shook like jelly, and sent at least 40 large buildings and 75 smaller ones crashing to the ground, while many others cracked or leaned over dangerously. Over 6000 people died in Mexico City (some reports suggest that the total was over 25 000), and

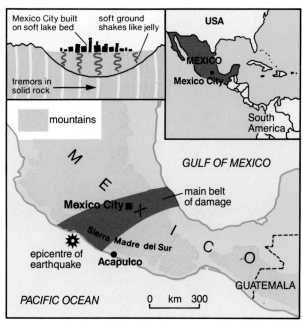

22 Mexico: the area affected by the 1985 earthquake

thousands more were injured and trapped under tonnes of fallen concrete. Blocks of flats in which people lived collapsed like cards, as did hotels, hospitals, and schools (see photograph 23). Other buildings swayed to and fro dangerously during the earthquake, and windows ex-

23 Earthquake damage, Mexico City 1985

ploded outwards showering people below with huge fragments of glass. Fire swept through many of the ruined buildings as gas pipes and electricity cables were cut. Roads, railways, and telephone links were cut, and airlines were told not to fly over the ruined city.

Between Mexico City and the Pacific coast, there were many other reports of damage. In one town a cathedral collapsed during a service, killing 26 people. At sea, four freighters and nineteen fishing trawlers were reported missing, believed sunk by tidal waves.

The next day, another earthquake, measuring 7.3 on the Richter scale, struck the city causing more buildings to collapse. This was followed by smaller earthquakes and tremors, called **aftershocks**.

As immediate rescue operation was launched involving expert rescue teams from Britain, France, West Germany and Switzerland as well as Mexico. Sniffer dogs, heat-sensitive equipment, and video-cameras were all used to search for those buried alive in the rubble (see photograph 24). Many spectacular rescues were shown on television at the time, but the most dramatic was the rescue of over 40 newly born babies from the ruins of a hospital. Some of them had been trapped for over a week.

Other help and money was flown in from all over the world. Tents for the thousands of homeless families, food, drugs and first-aid equipment, are some examples of aid sent from international organisations and other countries to a disaster area like this. You will find out more about this in Chapter 6.

Disastrous earthquakes like the one in Mexico happen regularly at plate boundaries, especially around the Pacific Ocean where some of the largest earthquakes in the world have occurred (see map 25).

24 Rescue work after the earthquake, Mexico City

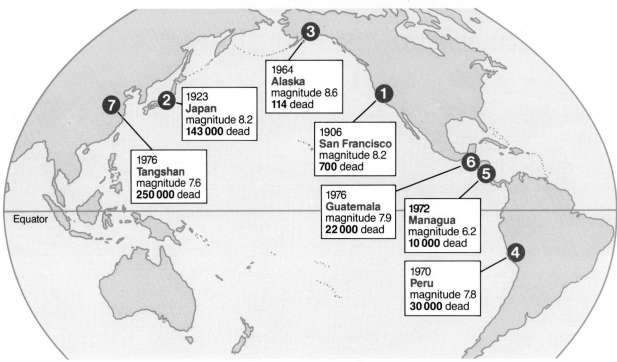

25 The Pacific earthquake belt

Core work

25 (a) Why was Mexico City so badly affected by the earthquake?
   (b) In your own words, describe what happened when the earthquake struck Mexico City.
26 (a) Imagine you are a rescue worker. Describe the exhausting work of digging for survivors, and the joy of finding someone alive.
   Or
   (b) Imagine you were in a block of flats when the earthquake struck. Describe what happened to you.
27 Look carefully at diagram 21.
   (a) Name the two plates that collided causing the Mexico earthquake in 1985.
   (b) What causes earthquakes to happen?
   (c) How are fold mountains like the Sierra Madre del Sur formed?
28 Make a table like the one on the right. In the spaces on the right, write sentences to show the effect of the earthquake on Mexico City and its people.

Extension work

29 Study map 25.
   (a) How many people have been killed this century by large earthquakes around the Pacific Ocean?
   (b) Compare maps 25 and 7. How many large earthquakes occurred near to plate boundaries?
30 Describe or draw labelled diagrams to show the three ways in which plates move.
31 Find out more about one of the serious earthquakes shown on map 25. Look for the information in your library or ask your teacher for help. Write down some details about it in your workbook.

Event	Effects
1. Earthquake shakes ground	1.
2. Buildings fall	2.
3. Gas pipes, electricity cables cut	3.
4. Fresh water and sewers damaged	4.
5. More earthquakes, aftershocks	5.
6. Rescue work takes place	6.
7. Aid from abroad	7.
8. Photographs, TV film goes abroad	8.

2 Nature shapes the landscape

Weathering

If a piece of clay is taken from the ground it can be pounded by machine or by hand, mixed with water and pressed and smoothed into a ball. A sculptor can then cut it, squeeze it and mould it into a new and recognisable shape (see diagram 1).

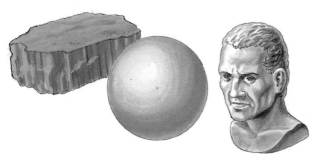

1 Modelling a piece of clay

The surface of the land is shaped in a similar way. Earth forces (see Chapter 1) push new rock up to the earth's surface by raising great blocks of rock as mountains and by volcanoes pouring lava on to the land.

The **agents of Earth sculpture** (running water, sea, ice, wind) then go to work on the new rock by cutting into it, carrying it or moving it, building on it and changing it into new shapes and forms. These agents are responsible for making the valleys, hills, gorges, beaches, caves and many other features we see around us.

However the new rock is *not* like the softened clay used by the sculptor. It is very difficult just to rub it away or even cut it with a saw or chisel. Sometimes it is difficult to even scratch the surface. How then can running water, waves and wind manage to change the shape of the rock?

Nature must first of all break the surface of a new area of rock (see diagram 2). Only then will pieces of rock be moved or carried away.

The surface layers of new rock are broken into pieces by the action of **weathering**. Here are some of the ways in which weathering can work.

2 A new landscape being weathered

- **Water from rainfall gets into cracks in the rock.** If the water freezes it expands by nearly 10% of its volume. The ice then wedges or pushes the crack open and weakens the rock. You can often see this happen with harling on a house. Water gets between the bricks and the layer of cement harling that covers them. When the water freezes, the harling is levered off the wall and gravity does the rest (see diagram 3).

3 How a house can be affected by weathering

- **Plant roots can also break up rocks.** Rootlets climb into tiny cracks and as the roots grow larger they push the rocks apart (see diagram 4).

4 How plants can cause weathering

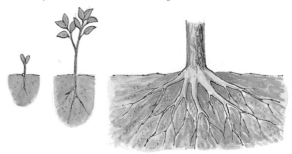

- **Rain is a very dilute acid which can dissolve some rock**. Also, when plants and bacteria die and decay, the decay produces acid. When these acids run over rocks and into cracks, they slowly dissolve some of the substances in the rock (see diagram 5).

5 (a) This is rock made up of many substances or minerals which have been pressed and cemented together

(b) Now some of the minerals have been dissolved so the rock looks rather like a piece of sponge

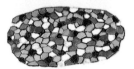

(c) The rock then easily collapses and the particles can be swept away

- **Rubbing and scraping can also wear away the surface of rocks** by removing particles. This is particularly true of softer rocks. Waves can pick up rough and angular pieces of rock. By moving them around so they rub against each other, rounded stones which are smaller than the original are produced (see diagram 6). Even people's feet can do this as you can see in diagram 7.

6 *From rock fragments to pebbles (the shaded part has been removed, making the rock smooth)*

7 *Stairs in an old stone building showing how they have been worn*

8 *This is a scene in a small valley where the rock has been weathered for thousands of years. The picture shows you some of the things that are now happening to the weathered rock.*

Core work

1 Make a list of all the ways given on pages 19 and 20 by which the surface of rock is weathered.
2 Study diagram 8. Try to find *four* things that have happened to the weathered rock. For example, at A the water from a thunderstorm has cut a small gulley by washing small rock particles into the stream.

Extension work

3 Try to find at least two examples of weathering near your home or school. Draw and describe them.
4 With the help of your teacher
   (a) try an ice expansion experiment with a plastic bottle;
   (b) use some vinegar to make limestone decay. Find out why this happens.

Erosion and deposition by running water

Loose fragments of rock, which have been broken down by weathering, often move downhill with gravity, or they may be washed down by rainwater, a process known as **rainwash**. These fragments also scratch and scrape the surface of the ground over which they pass, and wear it away. This whole process of wearing away the landscape is called **erosion**.

There are many agents of erosion which play a part in wearing away the landscape. The main agent of erosion in most countries today, however, is **running water**.

In the mountains

A heavy rainstorm, or water from melting snow, washes many of the smaller fragments down the mountainside. As the water runs downhill, armed with these sharp pieces of rock as 'tools', it gradually carves out small channels (or gullies) in the mountainside.

Lower down the mountainside, these gullies become small valleys which have streams flowing in them all the time. The force of the water in these streams is often powerful enough to move along large stones and even small boulders (see photograph 9). This usually happens when the streams flood after heavy rain, or when the snow is melting.

The stones carried along by the streams continue to wear away the rock along the sides and the bed of the stream. As a result, most valleys become larger lower down the mountainsides. At the same time, the stones and gravel carried along by the stream are themselves worn away as they roll along the bed of the stream and as they collide with each other. The size of the material carried along by a stream therefore usually gets smaller towards the lower ground.

Although fast-flowing streams carry out much erosion, not all of the work of widening and deepening a valley is carried out by the stream itself. In places where the weather is often wet, such as Britain, other things such as weathering and rainwash act on the valley sides.

All of these processes acting together usually form valleys which are V-shaped in cross-section. As the streams continue downhill, they are joined by smaller streams called **tributaries**. These add water to the main stream which eventually becomes large enough to be called a river.

In some parts of the world, where the weather is much drier, the action of rainwash on the valley sides is much less. Although the rivers which pass through these areas continue to cut downwards and deepen the valleys, the valley sides remain largely untouched. Deep steep-sided valleys called **canyons** are therefore formed (see photograph 10).

9 A mountain stream

10 A canyon

Similar features are also formed where rivers are cutting through very hard rock. Although the river is powerful enough to keep on deepening its valley, weathering and rainwash are often not powerful enough to remove the hard rock of the valley sides. Again, deep steep-sided valleys are formed, this time called **gorges**.

Core work

5 (a) Explain what is meant by
   (i) weathering, (ii) erosion.
 (b) What is the main difference between weathering and erosion?
6 Why do streams usually flow quickly in mountain areas?
7 (a) What is (i) a tributary, (ii) a gorge, (iii) a canyon, (iv) rainwash?
 (b) What are the main things which wear away (i) the channel of a river or stream, (ii) the sides of the valley?
8 (a) When is the stream shown in photograph 9 likely to move larger stones?
 (b) Why are valleys usually larger lower down the mountain than higher up?
 (c) Why is the material carried by a stream usually smaller in size lower down the mountainside?

Extension work

9 What are the 'tools of erosion' of a river or stream?
10 Use a dictionary to find out what is meant by the source of a river.
11 (a) How is the valley in photograph 10 different from the river valley shown in photograph 9?
 (b) Why are the shapes of these two river valleys so different?
12 Use your library to find out more about some of the processes of weathering mentioned earlier. Draw diagrams and write a few sentences to explain what happens in each of these processes.
13 Study photograph 10 which shows a river valley in the dry south-west of the USA. How can you tell that this is a drier area than Britain? Check this in an atlas.

11 Erosion and deposition by a river

In the valleys

Rivers continue to flow down from the hills until they eventually reach areas where the ground is lower and the slopes are much more gentle than in the mountains. As a result, they now flow much more slowly and so they are no longer able to carry some of the larger material along. These stones and pebbles are therefore dropped, or **deposited**, along the sides or on the bed of the river. This is often seen most clearly at a bend on the river, as in photograph 11. Here, small stones are being deposited on the inside of the bend, where the river flows more slowly. If you look at the outside of the bend, however, you will see that the river bank is still being cut away. This is because the river flows more quickly on the outside of the bend.

Near the sea

As the river nears the sea, it can no longer continue to deepen its valleys because it is nearing sea level. The river will therefore use its energy to cut away the outsides of the river bends, eroding sideways rather than downwards.

While this is going on, weathering and rainwash continue to act on the valley sides. The slopes on either side of the river become smooth and much more gentle. A typical river valley near the sea is therefore wide, with gently sloping valley sides. Large bends called **meanders** often develop on the river, which swings from side to side of the valley (see photograph 12).

12 A river meander

The land on either side of the river is usually fairly flat and so it is easily flooded if the level of the river rises. This flat area on either side of the river is therefore called the **flood plain**.

When it reaches sea level, the river becomes **tidal**. This means that sea water will flood into the lower part of the river at high tide, raising its level. At low tide, the sea water retreats and the river level drops again, often leaving sand and mud banks on either side of the river. The part of the river which is affected by the action of the tides in this way is called the **estuary**.

The actual shape of the river mouth varies from river to river. It depends on several things such as the size of the river, the amount of material that it carries, the action of the tides, and the currents off the shore. Sometimes at the mouth of a river there may be a **delta**. As the river water enters the sea, it suddenly slows down. The sands and gravels it was carrying are deposited and a fan-shaped feature called a delta is formed (see photograph 13).

13 The delta of a river

Core work

14 (a) Why do rivers usually flow more slowly in the lowlands than in the mountains?
   (b) What happens to the material being carried by rivers when they slow down?
15 (a) Describe the river bank on the left of photograph 11.
   (b) Describe the river bank on the right of photograph 11.
16 Explain what each of the following features is, and how it is formed: meander, flood plain, estuary, delta.
17 What do you think is meant by the 'tidal limit' of a river.

Extension work

18 Make a large sketch of photograph 12. Use arrows to mark and label the following features on your sketch:
   valley sides; the flood plain; meander; erosion taking place; deposition taking place
19 Use your library to find out more about how rivers meander and **ox-bow** lakes develop.
20 Why are sand and mud banks seen in estuaries at some times but not at others?
21 (a) Use your atlas to find different types of river mouth.
   (b) How would you describe the mouths of the following rivers: Thames, Clyde, Nile, Mississippi?

Summary

The course of a river and its tributaries, from its source to its mouth, is called a **river system**. In the last few pages, we have looked at some of the more important features which may be found in a large river system. These are summarised in diagram 14 below. However, it is not only in large river systems that such features are found.

A small beach stream, or a tiny stream flowing down the side of a spoil heap or sand quarry will often show some of the typical features of a river system on a smaller scale. It is in fact a complete river system in miniature. Look out for these around you: on holiday at the coast, or even locally in muddy areas, in parks, wasteland, or at the roadside.

Finally, there is one other agent of change which we cannot ignore in our study of river systems: *people*. Even high up in the mountains, people are helping to change the landscape.

If we look back at photograph 10, it is difficult to imagine how people could have much of an effect on an area like this. In photograph 15, however, we see that a dam has been built across this canyon. A large lake has been formed behind this dam to act as a reservoir. As dams are built to control and regulate the flow of water, less erosion is now carried out by the river below the dam, especially during the times when the river used to flood.

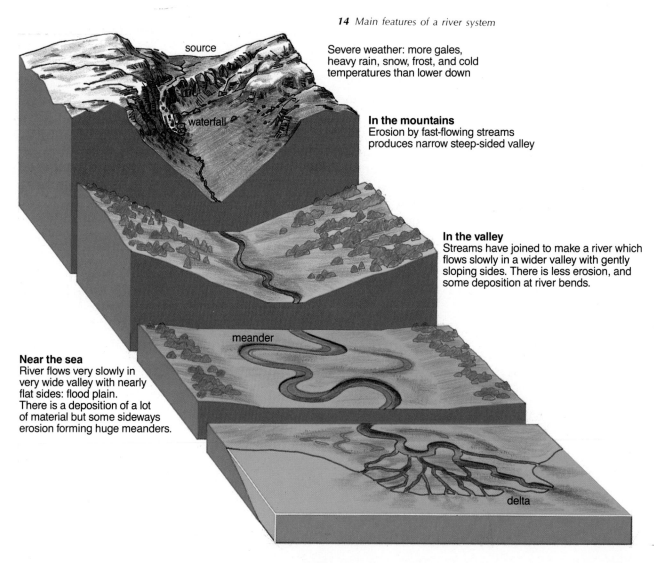

14 Main features of a river system

Severe weather: more gales, heavy rain, snow, frost, and cold temperatures than lower down

In the mountains
Erosion by fast-flowing streams produces narrow steep-sided valley

In the valley
Streams have joined to make a river which flows slowly in a wider valley with gently sloping sides. There is less erosion, and some deposition at river bends.

Near the sea
River flows very slowly in very wide valley with nearly flat sides: flood plain.
There is a deposition of a lot of material but some sideways erosion forming huge meanders.

15 A dam across a canyon

Some of the water from the reservoir is carried by canal to irrigate new areas of farmland in the desert nearby. In these areas too, people are changing the landscape.

Core work

22 (a) How has the landscape shown in photograph 15 been changed?
   (b) Why is this a good site to build a dam?
   (c) Why might a dam be needed in this area?
   (d) Apart from irrigation, what else can the water behind the dam be used for?

Extension work

23 (a) If you can find an example of a miniature river system in your home area, make a labelled sketch of some of the features that you can identify.
   (b) Some of these features can be produced experimentally in a laboratory sand tray or in a sand pit. If you can carry out any experiments, make sketches of your results.
24 Write a summary of the main features of a river system using the following headings; In the mountains; In the valley; Near the sea.

Erosion by the sea

Another important agent of change in the landscape is the action of the **sea** upon the land. Like running water, the sea erodes and deposits material, but only in coastal areas. Map 16 shows part of the east coast of Yorkshire in England. Here erosion is rapidly removing the soft clays and sands which make up the coastline. In some areas, the coastline has been retreating at an average rate of 2.5 metres per year. During a severe storm, however, erosion is much more rapid. The giant waves caused by the 'Great Gale' of 31 January 1953 cut back some areas of coastline by over 10 metres in a single night, causing tremendous damage. Even less severe gales cause considerable damage. Scenes like photograph 17 are still common where the rock beneath buildings at a cliff edge is worn away by the waves. The buildings eventually collapse on to the beach below the cliffs. As map 16 shows, over thirty villages are known to have disappeared into the sea along this stretch of coastline since Roman times.

The sight of waves breaking along the coastline shown in photograph 18 gives us some idea of the power of the sea. However, most of the erosion is carried out, not by the waves themselves, but by the stones and pebbles which they carry along. These are the 'tools of erosion', which the waves fling against the shoreline.

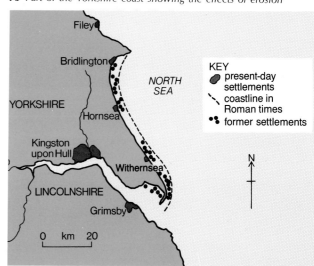

16 Part of the Yorkshire coast showing the effects of erosion

17 How a house can be undermined because of erosion

18 Waves breaking on rocks

There are three main ways in which the waves attack the coast, and they are summarised in diagram 19. Gradually, the coastline retreats as erosion continues and cliffs are formed. The speed at which the coastline is eroded will depend largely on the force of the waves, the amount of material which they carry and on the hardness of the rock which makes up the shoreline.

Waves are caused by the friction of the wind on the surface of the sea. They take some time to build up, however, and so large waves only form where strong winds blow over large areas of open sea. In Britain, for example, the Atlantic coast is most affected by wave action. However, other areas of coastline are also affected, as we saw earlier.

Erosion by waves does not take place at the same rate all along a stretch of coastline. Any joints or weak points in the rock are attacked first, so erosion is most rapid at these points. As they are enlarged, caves and arches are often formed (see diagram 20).

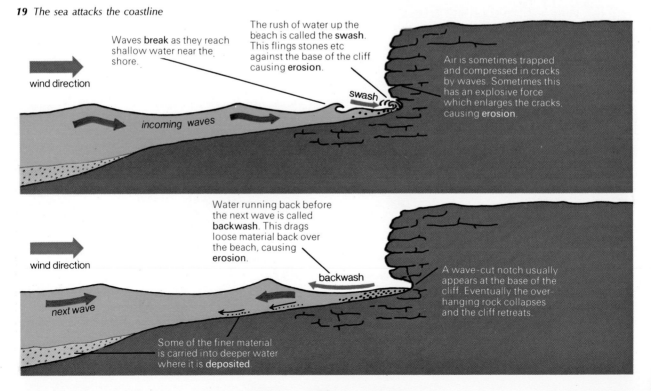

19 The sea attacks the coastline

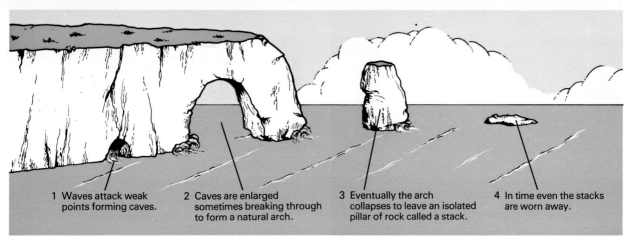

1 Waves attack weak points forming caves.
2 Caves are enlarged sometimes breaking through to form a natural arch.
3 Eventually the arch collapses to leave an isolated pillar of rock called a stack.
4 In time even the stacks are worn away.

20 Features found along a coastline of erosion

Core work

25 (a) Describe the ways in which waves may attack the coastline.
   (b) What are the main things which affect the speed with which the coastline is eroded?
   (c) How are waves caused?

Extension work

26 Study an atlas map showing the British Isles and the North Atlantic. Why should the Atlantic coast of Britain be most affected by wave action?

Deposition by the sea

Since waves are caused by the wind they approach the coastline from the direction of the prevailing winds. Quite often, therefore, the waves approach the coastline at an angle. This has the effect of moving material along the shoreline (see diagram 21). This type of movement is called **longshore drift**.

Sometimes this results in material being deposited across the mouth of a river or a bay. An area of calm sheltered water is therefore formed, protected from the open sea by a **sandbar** or a **sandspit** (see photograph 22). Deposition often occurs in this area of sheltered

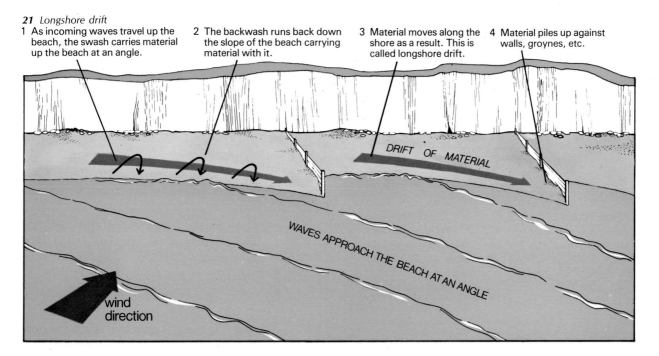

21 Longshore drift
1 As incoming waves travel up the beach, the swash carries material up the beach at an angle.
2 The backwash runs back down the slope of the beach carrying material with it.
3 Material moves along the shore as a result. This is called longshore drift.
4 Material piles up against walls, groynes, etc.

22 A sandbar or sandspit

water, and marsh vegetation grows as the water becomes shallower. The roots of the vegetation then help to trap sand and silt, making the water even shallower. Gradually the marsh dries out to form a new area of land.

Map 23 shows part of the English Channel coast between Kent and Sussex. Here deposition and the formation of new land have changed the coastline, and people have been affected as a result. After the Norman invasion of England in 1066, the small ports along this stretch of the coastline prospered as trade across the Channel grew. Gradually however, longshore drift built up a huge sand and shingle bar right across the bay and the ports were cut off from the open sea. Eventually the area behind the bar silted up and became new land. Today it is an important farming area which still bears the name of the Romney Marsh.

23 Changes along the Channel coast showing how deposition has altered the shape of the land

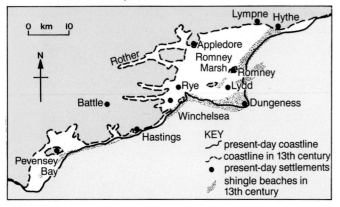

Deposition has continued to form the large shingle feature known as Dungeness. Although some parts of the marsh were artifically drained to form new farmland, deposition by the sea was the most important process which changed the coastline and the way of life of people living in this area.

Another kind of beach deposit is called a **raised beach** (see diagram 24). People built their houses on raised beaches because they had become covered with sandy soils which were suitable for growing crops. In mountainous areas of Scotland they are often the only places where cultivation is possible.

Behind raised beaches there are often lines of old sea cliffs: a reminder that the sea has not always stood at the level at which it stands today. Sometimes it has been higher and sometimes it has been lower.

24 Raised beach

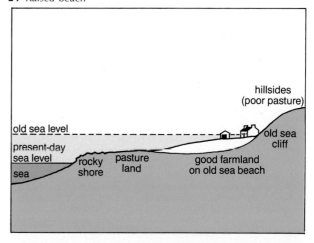

Core work

27 (a) What is longshore drift?
   (b) Draw a simple diagram to show why material moves along the shore in this way.
   (c) How does marsh vegetation help in the formation of new areas of dry land?

28 (a) What is a raised beach?
   (b) Why are people often attracted to raised beaches, especially in areas like the one shown in diagram 24?

The work of moving ice

In some parts of Europe – the mountains of Wales, Scotland and the Alps in Switzerland, for example – there are valleys like the one shown in photograph 25. The shape of these valleys is not like the V-shaped valleys formed by rivers and streams. Instead they are wide, deep, steep-sided and U-shaped in cross-section. Many of these valleys only have small streams flowing in them, too small in fact to have formed valleys of this size.

We can learn about the formation of U-shaped valleys if we look at some of the colder parts of the world.

Photograph 26 was taken high in the Swiss Alps where the weather is cold all year round. The snow seldom melts very much and it builds up to great depths in the valleys and hollows. When this happens, the weight of the snow causes the snow at the bottom to turn to ice.

If the snow continues to build up in the valleys and the hollows, the ice at the bottom behaves in an unusual way. It starts to 'flow', as molten plastic might do, carrying the snow and ice above it along too. In this way, a slow-moving 'river' of ice called a **glacier** is formed (see photograph 26 and diagram 28).

25 *A U-shaped valley in Switzerland*

26 *A glaciated valley in Switzerland*

The low temperatures in these areas also have an important effect on the landscape. Frost and ice break off small pieces of rock and the broken fragments fall on to the glacier below. There they become embedded in the ice or fall down large cracks called crevasses. Photograph 27 shows in close up the sharp fragments of rock which have reached the bottom of the glacier. The ice which you can see at the top is moving slowly from right to left. The rock fragments are dragged along by the moving ice and they scratch and scrape away at the rock surface over which they are passing like sandpaper or a nail file. Moving ice is therefore a powerful **agent of erosion**.

27 *Erosion under the ice*

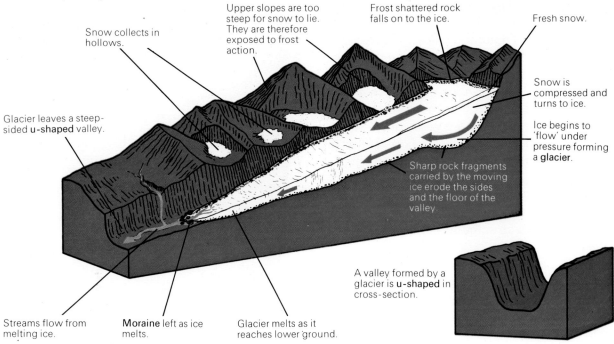

28 Erosion by moving ice

Erosion by moving ice is similar to erosion by running water in some ways. In both, for example, rock fragments and stones act as the tools of erosion. The shape of the valleys formed by each process is different, however.

29 Annotated sketch of photograph 30

30 Glacial scenery in Greenland

31 Glacial deposition

We now know that, on three occasions in the last million years, the weather in Britain was much colder than it is today. Snow fields and ice caps formed in the mountains, and glaciers moved out from them. Valleys like the one shown in photograph 25 were formed by glaciers during these cold periods which we call the **Ice Ages**. Most of Britain north of a line drawn from the Severn Estuary to the Thames Estuary was covered by snow and ice at some time during the Ice Ages.

In some parts of the world the Ice Ages lasted much longer than in Britain and spectacular ice-carved landscapes were formed (see photograph 30). Similar features can also be found in Britain, but usually they are on a smaller scale.

Glaciers do not only erode the landscape, however. When the ice stops moving forward and begins to melt, the material that it carries is deposited. Photograph 31 shows a valley high in the Alps from which the ice has recently retreated. The mounds of material in the foreground were left behind as the glacier melted. This material is called **moraine**.

People and ice

Glacial valleys are very important to people who live in mountainous areas. They are often the only places where the land is low enough or flat enough to build settlements. The deep valleys may also be useful in other ways. They can be dammed to form reservoirs (see photographs 15, 9 in Chapter 4, and 15 in Chapter 7). Some reservoirs provide water for hydro-electric power, while others provide large towns and cities with their water supply. For example, in Britain, Manchester gets much of its water from the Lake District, while Liverpool and Birmingham receive theirs from the Welsh Mountains.

Core work

29 (a) What is a glacier?
   (b) Draw a simple sketch to show how a glacier is formed.
   (c) How is the valley shown in photograph 25 different from the river valley shown in photograph 9?
30 (a) What do we mean by the Ice Ages?
   (b) How can you tell that the valley shown in photograph 25 was cut by a glacier?
   (c) In which direction do you think the glacier was moving? Why?
   (d) What is moraine?
   (e) How can you tell that the material shown in photograph 31 was not left behind by a stream or a river?

Extension work

31 Make a list of the ways in which glacial valleys may be of use to people.
32 Find a detailed map of each of the following areas in your atlas: the Lake District; the Scottish Highlands; the Alps.
   (a) Write down the names of two valleys from each area and the name of a town in each of these valleys.
   (b) How many of the settlements in *each* of these areas are found in the valleys? Choose your answer from: none; few; most; all.
33 Some parts of Britain which have an ice-carved landscape attract millions of tourists every year.
   (a) Try to find out why tourists are attracted to areas such as the Lake District, Snowdonia, the Cairngorms.
   (b) Make up a tourist brochure called 'A Mountain Holiday' to attract people to one of these areas.

33 Desert erosion

The work of the wind

If you have walked into a strong wind on a dry day, you may have been bothered by dust and fine grit blowing into your face and your eyes. Before the wind can move material along in this way, the material must be lying loose on the surface. If the ground is covered by vegetation, the roots will hold the soil together so the wind will not move it. The work of the wind is therefore most active in places where the surface is not protected by vegetation: for example, in newly ploughed fields, on beaches or in deserts.

Material blown by the wind is not usually moved far before it is deposited again. However, on beaches where strong winds are coming from the sea the finer material is blown into sand hills called **dunes**.

Dunes are not permanent features, because the sand begins to move again when a strong wind blows. Usually the dunes creep forward slowly in the direction of the prevailing wind. However, storms may cause much more rapid movement (see diagram 32). Sand dunes 20–30 metres high are found in Death Valley, a desert area in California. Here the wind is acting as an **agent of deposition**. Every time a strong wind blows, the shape and the position of the sand

32 The formation of sand dunes

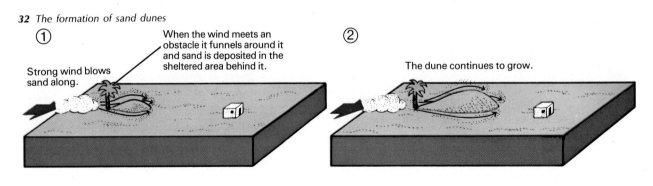

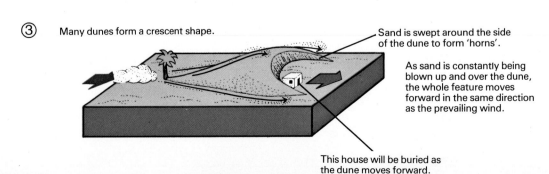

dunes is likely to change. Areas of dunes which shift in this way are called **sand seas**. Not all deserts are made up of large areas of sand dunes however. Photograph 33 shows an entirely different desert landscape. Here the wind has acted mainly as an **agent of erosion**. As fine grains of sand are blown along close to the ground, they have a 'sand blasting' effect on the rocks. If the rocks are very hard and unjointed, they may simply be polished and rounded. If there are joints and weaknesses in the rocks however, they can be carved out to form some unusual shapes.

In some desert areas, sand blasting can cause people problems when the bases of fences and telegraph poles are cut away. To avoid this they are often protected by metal shields.

To prevent sand dunes moving, tough **marram grass** is often planted. Once the long roots begin to anchor the sand, coniferous trees can be planted. Their long shallow roots fix the dunes more permanently. In this way the danger is removed and at the same time useful timber can be produced.

The wind can also threaten farmland. In East Anglia in England, for example, it is thought to blow away about 2 centimetres of soil from farmland each year. To cut down this soil erosion, people often plant long lines of trees. These **shelter belts** help to reduce soil erosion by breaking the force of the wind.

Core work

34 (a) Make a list of the places in which you have seen the wind moving loose material about.
   (b) Why are you most likely to see this movement on a dry day?
   (c) Why is wind action less active in places where there is a lot of vegetation?
35 Draw a simple diagram to explain how sand dunes are formed (see diagram 32).
36 (a) How does the action of the wind affect people?
   (b) How do people try to reduce the movement of sand and soil by the wind?

Landscapes shaped by climate

As we have seen, earth forces, the action of weathering, erosion and deposition all shape the landscape in different ways. Nature affects the landscape in other ways too, however. Areas of the world can have different landscapes because their **climates** (average weather conditions) are vastly different. In the rest of this chapter, we shall study three **natural regions** to see how extreme climates affect the landscape. These areas have hardly any people living in them and are sometimes called **empty** or **difficult areas** because they are difficult to live in. Map 1 on page 38 shows where these areas can be found in the world. Most people avoid living in these areas because the climate is very uncomfortable. There is usually too much or too little heat or water, as we shall see below.

Cold deserts or tundra areas

In Greenland and parts of northern Canada and northern USSR, it is very cold for most months of the year and the thermometer rises above freezing point for a few months only. These areas have a cold desert or **tundra** climate: very few plants will grow and few types of animal live there.

Graph 34 shows an example of the climate of the tundra. Here only between May and September are the temperatures likely to rise above freezing point. For the rest of the year the landscape is covered by a thick layer of snow and ice (see photograph 35).

34 Climate graph for a tundra area

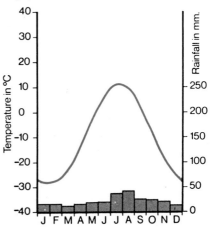

35 The tundra in winter

Plants will not grow until the temperature rises above 6 °C. This temperature is called the **growing point**. Only plants which can complete their yearly life cycle in a few weeks can therefore grow in the tundra. These include mosses, lichens and a few small flowering plants. For a few short weeks, the tundra can be a mass of colour (see photograph 36), as these plants struggle to flower before the long winter sets in.

Not surprisingly, few people live in these harsh areas. Those who do have had to adapt to life under severe climatic conditions.

Core work

37 (a) Write down ten words or phrases which describe the tundra.
   (b) Name three countries with large areas of tundra.
38 What is climate?

Extension work

39 Compare the tundra climate graph 34 with a climate graph of a place in Britain. What are the main differences?
40 Find out about ways in which the Lapps, Inuit or Samoyeds have adapted to life in the tundra.

36 The tundra in summer

Deserts that are hot or warm

In hot deserts the climate is so dry that few plants will grow. Usually an area is called a desert if on average less than 250 mm of rain falls each year. Graph 37 shows a typical hot desert climate. In warm deserts the temperature would be much lower but there is still very little rainfall.

The kind of picture people have of a hot desert is sand dunes, oases and camel trains. But deserts can be rocky or sandy (see photographs 33 and 38), and can become very beautiful places after rain (see photograph 39).

As it seldom rains in the desert, any plants which do grow have to be able to survive long periods of drought. Some, like the cactus, can store water in fleshy leaves or stems. Others grow quickly after a shower of rain, produce flowers and seeds, and then wither and die after a few days.

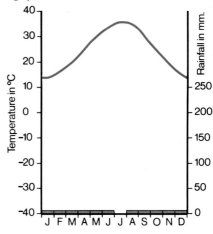

37 Climate graph for a hot desert area

38 Sandy desert

39 Desert after rain

Core work

41 (a) Write down ten words or phrases that describe hot or warm deserts.
   (b) Why are warm or hot deserts not easy places for people, plants or animals to live?
   (c) Can you think of any reason why people might go to live in a hot desert?
   (d) Describe the main differences between photographs 38 and 39.
   (e) Why do you think one desert in California is called Death Valley?

Extension work

42 Find and read a story of a desert journey or survival in a desert.

Hot wet areas or tropical rain forest

We have found that few people live in the tundra because temperatures are so low that very little can survive. Few people live in deserts because they are too dry to support much life. Unlike these areas, **tropical rain forest** has high temperatures and heavy rain throughout the year. This provides ideal conditions for plant growth, and so thick forests cover the area. These areas often contain many unusual plants and trees (see photograph 40).

The trees often grow so close together that their branches form a continuous layer called a **canopy** (see diagram 41). Under this canopy, the heat and moisture combine to produce a very humid, hot-house type of atmosphere. Europeans in particular usually find this type of atmosphere unpleasant to live and work in.

Although many people find these areas unattractive, insects like them and diseases such as malaria and yellow fever are common.

The heavy rain can also be a problem to people. Flooding often occurs, and if the vegetation is cleared away to allow farming to take place, minerals are soon washed away in the wettest areas. Movement through the thick forest

40 Tropical rain forest in the Amazon Basin

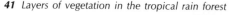
41 Layers of vegetation in the tropical rain forest

Tallest trees called **forest giants** may be over 100m high.

Branches of the middle layer join to form a canopy.

Beneath the **canopy** the air is still. Very humid hot-house-type conditions develop.

Very little light reaches the forest floor so little undergrowth grows. A thick layer of dead leaves, branches etc. forms.

Many varieties of plants grow in these hot-house conditions, such as giant tree ferns, and many climbing plants and creepers.

Tallest trees have **prop** roots.

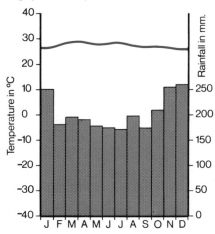

42 Climate graph for a tropical rain forest area

is usually difficult and rivers are often the only means of transport.

Despite all of these difficulties, there are areas of tropical rain forests which have many inhabitants. One example is Java in Indonesia, where the combination of the fertile volcanic soils, the rain and the great heat allow rice, bananas and other crops to grow. Another example is the coast of West Africa where many farmers grow cocoa and other plantation crops. However, there are many other areas where the difficulties have been too great in the past for much development to take place. The Amazon Basin of South America is an example and we will look at this area in Chapter 3.

Other difficult regions

43 A steep valley in Tibet

Core work

43 (a) Describe the tropical rain forest area shown in photograph 40.
   (b) Make your own labelled sketch of the forest using diagram 41.
   (c) Use graph 42 to describe the climate of the area.
   (d) Why is this area a difficult one for people to live in comfortably?

Core work

44 Look at photograph 43.
   (a) Work out reasons why this area has problems for people wishing to live there (farming the land, building roads and railways, and building towns and villages).
   (b) Photograph 43 shows an area in Tibet. Use your atlas to find this place.

3 People begin to change the landscape

Introduction

Chapters 1 and 2 described how the earth's surface was created and then its shape changed by nature's forces. They also showed how patterns of weather continue to affect the land and how it can be used. This affects how many people can make a living from the land.

People have not always lived in and used all parts of the world. Even today, some parts are more useful and crowded than others. This chapter will look at one large area of land – North America – which was relatively empty until 200 years ago. It will describe how and why different groups of people moved across North America and settled in different areas at different times. It will look at the way these people changed the natural landscape and the lives of others living there.

Changes go on today, often faster and greater than ever before, yet old ways of life survive in some areas. This chapter will help you find out why. It will also show you that similar changes have happened or are happening, but now with much greater speed, in other parts of the world.

To begin with, we shall look at two maps that show which areas are difficult for people to use and live in, and where the crowded and empty areas of the world are.

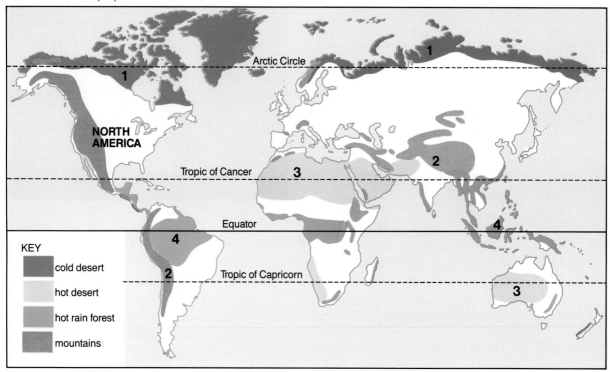

1 Difficult areas for people to use and live in

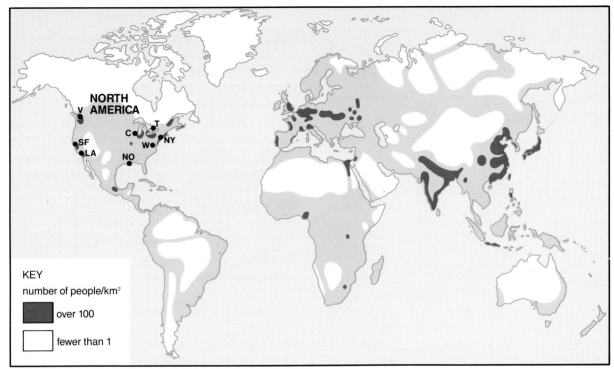

2 Crowded and empty parts of the world

Core work

1 (a) Copy and complete Table 1 in your workbook, using information in maps 1 and 2 (one example is done for you).
  (b) For each of the difficult areas, write down **one** reason why it would be difficult for people to live there.
  (c) Use a library, or resources that your teacher gives you, to try to find out about **two** difficult areas and how the people who live there make a living. Write down your findings and give them as a report to your group or teacher.

2 (a) Look at North America on map 1. Use information from map 1 to complete this paragraph in your workbook.
  Much of Canada, the country which occupies the n_____ of N_____ A_____ is made up of c_____ d_____. There is h_____ d_____ in the south-west of the continent and a wide belt of mountains runs from north to s_____ down the west side of the continent.
  (b) Map 2 shows some important American cities. Use an atlas to find their names. Write them down.
  (c) Which cities in your list stand on or near areas which are difficult for people to use and live in?

Table 1 Difficult areas

Area	Why is the area difficult to live in?	Do many or few people live there?
1	Cold desert	Few people
2		
3		
4		

39

NORTH AMERICA

In the great spaces of North America, nature created both difficult and easy areas to farm and work. There were many kinds of difficulties and many different resources for people to develop. Let us now look at how the Indians, the earliest settlers, began to change the natural landscapes they found there.

The first inhabitants

Map 4 shows where the first inhabitants of North America came from (around 12 000 BC), and where they settled.

3 North America

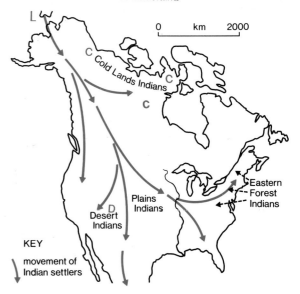

4 North America: the first inhabitants

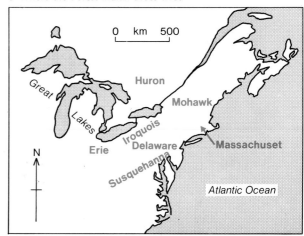

5 Where the Forest Indian tribes lived

Extension work

3 Use an atlas map which shows you the different types of landscape in North America, as well as maps 1, 2 and 3, to help you do this work.
  (a) Which of the following descriptions fits the landscape at places 1–10 on map 3?
      High mountains and deserts
      Fertile plains and rivers
      Cold high mountains
      Tundra barren land
      Hills and fertile valleys
  (b) Which parts would be easiest to farm? Why?
  (c) What is the reason for the crowded and empty parts of North America being where they are?

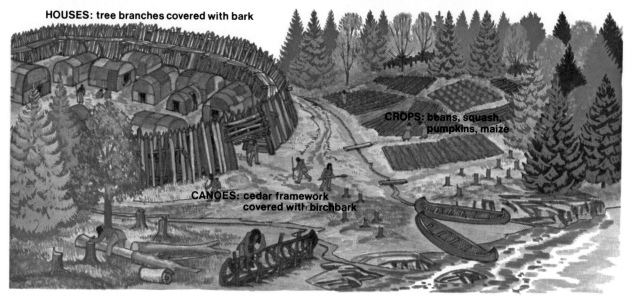

6 A Forest Indian village

The Forest Indians

As you can see in maps 4 and 5, the Forest Indians, such as the Huron and Mohawk tribes, lived mainly in the east of the continent. Their home was a vast forest which stretched inland for hundreds of kilometres. The rivers were rich in fish, and the woods offered plentiful supplies of game for hunting. Communication was easiest by means of canoes, so the Indians mostly lived in stockaded villages along the banks of the largest rivers. Apart from hunting and fishing, they knew about farming and cleared the forest to use for growing crops (see picture 6).

Core work

4 (a) In what ways did the Forest Indians make use of wood (see picture 6)?
  (b) Why do you think stockaded villages were necessary?
  (c) Describe the work and life of the village in picture 6.

Extension work

5 (a) Look at map 4. When the Indians arrived, there was a 'landbridge' at L. Why do you think it disappeared?
  (b) Can you suggest reasons for some Indian tribes moving into difficult areas like C and D?

The Plains Indians

Indians also settled on the Plains, the great rolling grasslands between the Mississippi River and the Western Mountains. To us, these are probably the best-known Indians. Tribal names such as Sioux, Dakota, Commanche, Cheyenne, Apache and Blackfoot featured in the many wars in the American West.

The Sioux are a typical tribe of the Plains. Their way of life revolved round two animals: the horse, which was introduced to North America by the Spanish; and buffalo which were once found in such large numbers grazing upon the great grassland that they were likened to a 'single moving carpet' on the Plains.

Unlike the Forest Indians, the Plains Indians did not live in permanent villages. They moved camp as the herds of buffalo moved in search of food and water.

7 Indians hunting buffalo

8 The uses of the buffalo

Extension work

7 (a) Use the scale on map 9 to work out the distance: (i) from north to south, (ii) from east to west of the Sioux hunting ground. How much bigger is this area than the UK?
 (b) Why did the Sioux need such a large hunting area?
 (c) Copy Table 2 into your workbook and fill in the columns. Use pictures 6 and 8 to help you.

Table 2

	Forest Indians	Sioux
Type of vegetation		
Type of house		
Materials used to make houses		
Ways of getting food		
Methods of transport		

Core work

6 (a) Study pictures 7 and 8 and list the way in which the Indians made use of buffalo.
 (b) Why do you think the buffalo was easy to hunt and kill?

People on the move

There are many reasons why people move from one place to another to live. Mostly, people move because something attracts them to a new place (**pull reasons**), or because a change occurs in the place where they are living which makes them want to move out (**push reasons**).

Here are some of the reasons behind the

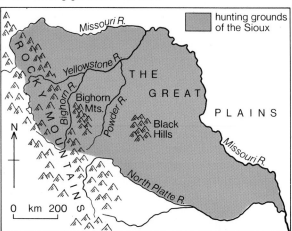

9 The hunting grounds of the Sioux

10 European settlers travelling to eastern America

movements of people into and inside North America.

Indian peoples: discovery of better hunting ground, too much competition from other tribes for game, poor climate, lack of animals left to hunt.

European peoples: search for gold, famine, religious troubles, need to trade, desire to explore, shortage of farming land, discovery of trapping areas, disease, harsh rulers (political oppression).

Core work

8 (a) Arrange each list of reasons (for Indians and Europeans) into two columns, one headed 'pull reasons', and the other headed 'push reasons'.
  (b) There is still a lot of movement between countries and within countries. List as many other pull and push reasons as you can think of to explain some of these movements. (Think about why people from Britain emigrate to Australia.)

European settlers arrive

When the first West European settlers arrived in North America in the 1500s they found a landscape very similar to the one they had left. It would have been impossible for them to have crossed any other ocean and found such a familiar land. The summers were hotter and the winters longer and more severe, but the type of farming they were used to could still be carried on.

By 1700, the French and English settlers were claiming land as their own although, when they had arrived, the Indians were already living there. In this respect, the ways of the white settlers were very different from those of the Indian. The Indians found it very hard to understand what was happening when the settlers wanted to buy their land. To the Indian, land belonged to everyone in the tribe. It was the right of all hunters to seek game wherever they pleased. Only when a large amount of land had been snatched from the Indians, their forests cleared, and their hunting ruined, did they begin to resist in the only way left to them: fighting.

Core work

9 (a) What did the settlers from Europe do for a living when they first arrived in North America?
  (b) Why did the European settlers find the east coast of North America relatively easy to settle in?
  (c) What did they do to the land? Why did they do it?
  (d) What problems did this create for the Indians?
  (e) Why did the two groups end up fighting?

Extension work

10 Using picture 6 on page 41 as your starting point, draw a picture to show what the area might have looked like after the European settlers moved in.

European settlers move west

It was fairly easy for the first settlers to move the Forest Indians off any land they wanted as they opened up the east coast. Gradually they moved west in search of more land.

Core work

11 Diagram 11 is a typical silhouette of the shape of the land across the middle of North America. Sketch this cross-section into your workbook and, using an atlas, name (a) mountain ranges 2, 5, 6, 7 (b) oceans 1, 8 and (c) rivers 3, 4.

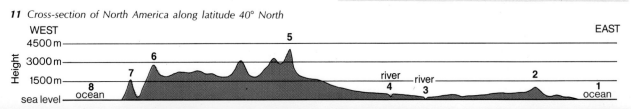

11 *Cross-section of North America along latitude 40° North*

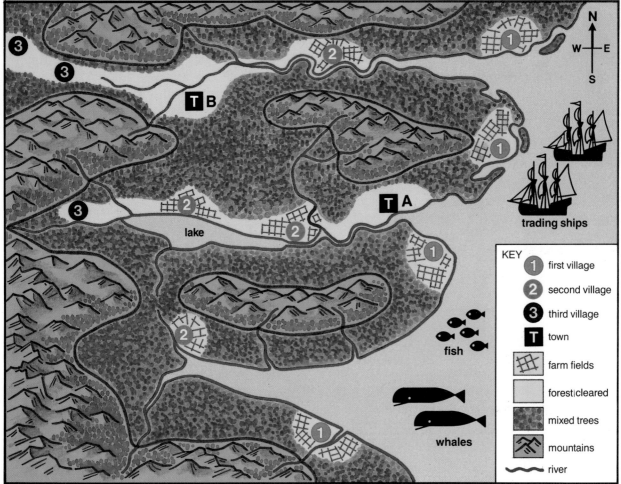

12 Map for quiz game

Quiz game

Make a tracing of map 12, or get a copy from your teacher. Look carefully at the map for clues to help you with this quiz. First villages are the ones settlers built when they first arrived. Second and third villages are the ones they built later.

(a) Only some of the first villages have been shown. Put in two more where you think they would probably have been.
(b) Write down one reason why the first villages were built where they were.
(c) Name two ways that the people in them might have earned a living.
(d) Look at where the second villages grew up. Write down two reasons why you think they grew up there.
(e) Draw in two more second villages in similar places.
(f) Name two ways villagers in them might have earned a living.
(g) Look for an area with no villages. Explain why none would be found there.
(h) Look for the third villages. Would they have been built before or after the second villages? Why do you think so?
(i) Imagine you lived in a third village. You have goods to send back to Europe. On your map use a colour to mark the route you would take to get to the coast. Why would you take that route?
(j) Look at town A. What reasons can you think of for it growing up where it did?
(k) Look at town B. What reasons can you think of for it growing up where it did?

13 The wagon-train rolls west

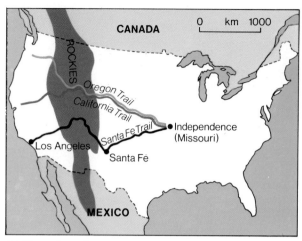
14 The trails to the west

European settlers come to the Great Plains

The quiz game on page 44 showed how European settlers in North America moved west quite easily by travelling along rivers or clearing forest for farming. When they reached the Great Plains, however, things became more difficult. Trees began to disappear and coarse grass took its place. This land was much harder to work with simple wooden ploughs. The climate also became harsher and more extreme.

Unlike the Forest Indians, the Plains Indians remained undisturbed by white settlers for many years. The hard conditions put off the settlers and it was not until they had a very good 'pull' reason that they ventured across the vast rolling plains. The pull reason was the 'promised land' of California in the furthest west where the land was fertile, the climate mild and where, in 1848, gold was discovered. News of the gold and the promise of a better life soon led to a flood of movement across the Plains to the west. In one year, over 50 000 people crossed the country (see picture 13 and map 14). As diagram 15 shows, the journeys were difficult and dangerous.

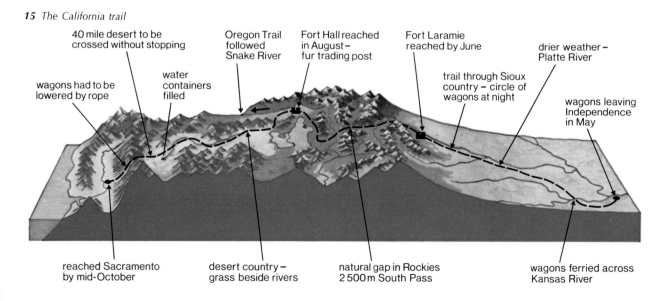
15 The California trail

Core work

12 (a) List the pull reasons which attracted people westwards.
   (b) The wagon trains carried all their own provisions except for fresh meat. Why do you think this was?
   (c) In one early journey west, only 45 out of 90 people reached California. Suggest reasons for this. (Look at diagram 15 for clues).
   (d) In your own words, describe the California trail (diagram 15). Mention some of the natural barriers met by the travellers and say how they overcame them.

Extension work

13 (a) Look at an atlas map showing the area crossed by the Santa Fé trail. Then try to draw your own map, adding present-day place names, and the names of rivers and mountains through which the old trail passed.
   (b) Which areas along the trail would have been pleasant to travel through? Why?
   (c) Which would have been dangerous? Why? (Diagram 15 gives clues for this question.)

16 Building the railroad west

17 Buffalo being shot by settlers for sport

End of an Indian way of life

The Sioux and other Plains Indians regarded this mass movement west with growing fear. The frightened buffalo were beginning to move away and the Indians soon started to fight to save their land and animals. But worse was to come when the railroads were built (pictures 16 and 17). The Indians' way of life became threatened, as railway gangs began to shoot huge numbers of buffalo for sport. This is what it seemed like to the Chief of the Sioux:

> 'He tried to count, but found this impossible because of the quarrelling of the buzzards. Even so he decided that there must be two hundred or more carcases there, stripped of hides worth three or five dollars in Kansas. The flesh left for the buzzards, wolves and ravens. The Sioux sat in dejected silence, staring at the slaughter. "Many buffalo" he said. "Why kill and not eat?"'

Core work

14 How would shooting huge numbers of buffalo destroy the Indians' way of life?
15 (a) Settlements began to grow up beside the railroads. Why?
   (b) How did the railroads make it possible for people to settle on the Plains and make a living?
   (c) Look at picture 16. Imagine you lived in a railroad settlement. Describe the kinds of buildings and activities you would find there.
16 Look at pictures 18a–d on page 47. List the changes that (a) Indians and (b) Europeans settlers made to the natural landscape of the Plains up till the end of the 1800s.

18 Changes in the Great Plains

A broken people

The Sioux and other Plains Indian tribes were slowly pushed further and further west into the driest plains and mountain areas which were of little use to the European settlers. Those who resisted were finally defeated and the survivors were herded into **reservations**. Here they were encouraged to take up farming and to live a settled way of life. Today there are 282 Indian reservations ranging in size from a few hectares to areas as big as England and Wales.

Table 3 Some features of present-day life on a reservation

1. Life expectancy is 64, compared with a national average of 70 years.
2. Average Indian family earnings are only £1125 a year, whereas the national average in America is £5941, and the average in Britain is £3094 (1977).
3. There are social problems such as alcoholism.
4. Indian children often spend only five and a half years in school.
5. Many earn money by making and selling Indian craft products to tourists.

Many Indians have left the reservations and moved to the cities, but the problems they find there are often as great. Read the following comments made by present-day Indians:

'You will forgive me if I tell you that my people were Americans thousands of years before your people were. The questions is not how you can Americanise us but how we can Americanise you.'

'You said you wanted to put us on reservations. I do not want them. I was born where the wind blew free and there was nothing to break the sun. I want to die there and not within these walls.'

'... It was a jungle when we got to Chicago ... Nobody would invite you up to his house. I didn't feel I was human up there.'

Core work

17 (a) Map 19 shows the main Indian reservations. Describe where they are, and try to explain why they were set up there. (Diagram 15 may help you.)
(b) Why do you think the Indians found it difficult to farm the reservations? (Clue: look back at map 1.)

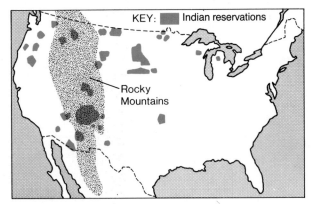

19 The main Indian reservations

The Plains area today

Since the building of railroads just over 100 years ago, the Plains have become a highly productive and prosperous farming area. This dramatic change to an area that was once thought to be a 'natural barrier' is due to many things, not least of which was the invention of barbed wire! The lack of trees meant that there was no wood for fencing, so early ranchers and sheepherders had a hard job keeping track of their animals. Barbed wire enabled them to make fields and boundaries and, as a result, they were helped to settle in one place.

Other inventions, such as wind pumps and stronger ploughs, also helped the early farmers to prosper and to push further west. As the numbers grew, so did competition for land.

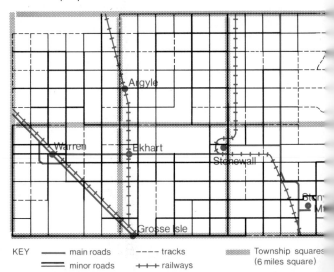

20 Township squares

21 The Plains today

Eventually, the settlers divided the land into 'township' squares, each one six miles square. Each township was then divided into thirty-six sections of one square mile and straight roads were built to connect one town with another. The result was a chequerboard landscape which still exists today (see diagram 20 and photograph 21).

Despite the present-day prosperity of the Plains, farmers still need to control their use of the land very carefully to overcome natural problems of the area. For example, they can
1. feed the soil by adding fertiliser;
2. rest the soil by planting different crops in rotation;
3. stop wind erosion by planting strips of trees and planting different crops between cereals to protect the soil;
4. irrigate the soil by pumping water from under the surface and by building dams.

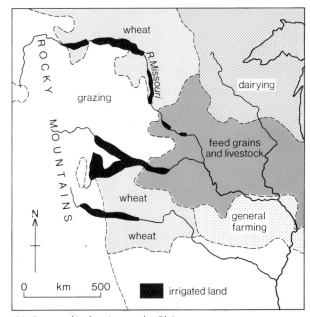

22 Present-day farming on the Plains

Core work

18 (a) Compare map 22 with map 9 on page 42. List the changes in the way land has been used.
   (b) Draw small sketches to show four ways in which the modern Plains farmer can control the land.

19 List the goods produced by modern Plains farmers (see map 22).

20 Describe the road pattern in diagram 20 and photograph 21. Why would such a pattern not be found where you live?

Surviving Indian lifestyles: the Hopi

Although today most of North America's landscape and lifestyle is based on the European settlers' traditions, Indian traditions have survived in some areas where reservations have been set up on traditional Indian lands.

The homelands of the Forest Indians and Plains Indians had natural features which could be put to good use. They were fairly easy areas to settle in. In other areas of the continent, for example the south-west, the climate is too dry for most crops and there is little natural vegetation. The south-west was therefore a **difficult area** to settle in. It was not attractive to farmers or hunters. Despite these natural disadvantages, however, Indians settled there and survived. We shall look at one of the tribes: the Hopi.

Photograph 23 and sketch 24 show the kind of landscape in which the Hopi lived and the ways in which they adapted to the surroundings. They built towns of stone and dried mud (adobe) on the sides of rock-sided mesas. The buildings were like apartment blocks, often rising to five storeys with as many as 800 rooms. Most ingenious of all was the Hopi Indians' ability to live by farming. Using springs at the foot of the mesas, they irrigated small fields and grew maize, cotton, beans and peppers.

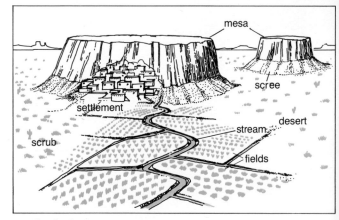

24 Sketch of a Hopi settlement

The Hopi, like other Indian tribes, now have to live in a reservation, but at least it is in the area where their ancestors grew up. Roads have been built through the area, bringing travellers and tourists. The tourist trade makes money for the Indians (see photograph 25), but most Hopi still earn their living from the land. Today, the villages (26) and farmland (27) are in many ways similar to the originals except that more crops are grown in greater variety, and fields are bigger. Changes in this case have developed the old way of life, rather than replaced it.

23 A mesa landscape in Utah

26 A Hopi village today

25 Indian performing for tourists

27 Irrigated fields in present-day Utah

Core work

21 (a) Write a short description of a Hopi village in the past (sketch 24 will help).
   (b) Do you think the villages were easily defended? Why?
22 (a) How did Hopi Indians make a living out of the desert in the past?
   (b) How can you tell that the area in photograph 27 is artificially watered (irrigated)?

Extension work

23 (a) Give reasons why the Hopi way of life has not changed much over hundreds of years.
   (b) Why is the area in which the Hopi Indians live a 'difficult' one?
24 What are the good and bad things about the tourist trade for the reservation Indian?
25 There are people who have come into the difficult south-west area. They test atom bombs, they dig for minerals, build dams and gamble. Find out more about these modern activities (pull factors) and the places in which they have happened.

A fast-changing mountain landscape: Appalachia

European settlers travelled through the valleys and passes of the Appalachian and Rocky mountains on their journey west. They were mainly farmers and, since the mountains were difficult to farm and make a living, few people settled there, so the area was left empty. In the 1800s, however, these areas were found to have fuels and metal ores which are very useful for the great industries which have grown up in North America. Mining these minerals has brought very great and rapid change to the natural landscape.

When people mine the surface of the earth to obtain a mineral, obvious and often massive changes are made to the earth's surface. We shall look at some of the changes that have happened in Appalachia, a coal mining area in the USA. Map 28 shows the area. Find it on an atlas map.

Many of the coal seams in this part of North America lie close to the surface and are often up to 2 metres thick. The seams are not greatly buckled by folding, nor are they broken by faulting. Diagram 29 shows how the seams lie in the mountains and the different ways in which they are mined. Notice that **strip** (or **open cast**)

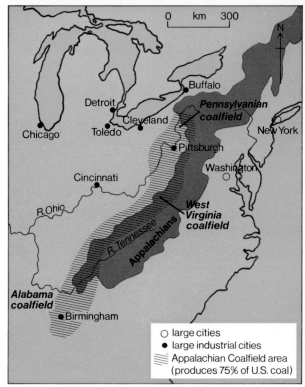

28 The Appalachian coalfield area

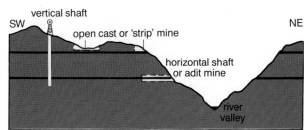

29 A typical cross-section through the Appalachian coalfield area

30 Strip mining in Appalachia

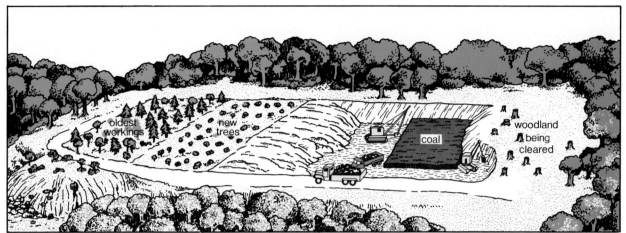

31 The 'cut and fill' method of mining

mining takes place at or near the surface of the ground.

In the past, strip miners simply bulldozed the soil away from a coal seam to get at the coal. After a while, rain and snow washed this soil and rubble down the valley sides, killing vegetation and clogging streams. Picture 30 shows evidence of how this type of mining changes the natural landscape.

Today, mining companies try to cut down these bad effects. One way is to try to repair the damage to the land surface (see picture 31). Strip mining methods are also used to obtain several other minerals. Iron ore and copper are both mined in this way in the Rocky Mountains.

Mineral resources are **non-renewable**: that is, they will eventually run out. Careful planning is therefore needed, not only to make sure that strip mining does not permanently spoil the landscape, but also to make sure that we do not use up our mineral resources too quickly.

Core work

26 Look at picture 30.
   (a) How do you think strip mining changes the land use in the area?
   (b) What effects do waste materials have on the landscape of the area?
   (c) How will the rivers be affected?
27 Study picture 31.
   (a) What types of machines are used to mine the coal and transport it within the workings? Why are such machines necessary for this kind of mining?
   (b) Why is mining generally easier in the valleys than on the ridges?

Extension work

28 How will settlements and people living in strip mining areas be affected by the work of mining companies who mine for coal?
29 The method of mining shown in picture 31 is called 'cut and fill'. Work out the mining sequence shown and describe it in your own words.
30 In any one week in the USA it is estimated that 500 hectares of land are ripped up by strip mining.
   (a) How many hectares of land are ripped up in one year?
   (b) List some of the problems this must cause.
   (c) Why is the 'cut and fill' method better than the old bulldozing method?

32 Port and central business area of New York

Fast-changing cities

The large city areas in North America have shown the greatest and most rapid changes to the natural landscape. Cities such as Pittsburg developed with the great iron and steel industry that grew up there using coal and iron ore mined from the nearby mountain area.

Later, great ports grew up along the east coast and on the Great Lakes. They imported raw materials to be used in the growing industries, and exported other products to Europe. Along with the wharves and harbours of the ports there grew up railway stockyards and great business centres with banks and finance houses, insurance offices and transport headquarters. Photograph 32 shows part of the port of New York and the central business area of the city.

In even more recent times the search for fuels has led to exploration under the sea bed of the Gulf of Mexico and the finding of oil. Oil came ashore in pipelines, and industry has grown along this shoreline which processes and uses the oil.

Core work

31 (a) Study photograph 32. Imagine you work in the office at the top of the building in the foreground. Describe the scene you might see from the window.
   (b) What businesses might be found in the tall buildings you can see?
   (c) Look at the road pattern. Why do the main roads follow the direction they do in the photograph?
   (d) Describe the street pattern. How is it different from a town or city you know?

33 Factories in Silicon Valley, California

34 San Diego

The most recent industrial growth is connected with electronics and computers. There is no need for such industries to be found near ores or fuels. Instead they are located where communications links are excellent and where small-scale components for the industry can be easily obtained. The landscape of such industries includes airports, motorways, smaller factories, little pollution and well laid-out green areas and pleasant paths. Photograph 33 shows such an area.

In America today, industrial growth and success, and the arrival of a new industrial technology which means machines do much of the work people used to do, mean that people have more time for leisure and more money to spend on it. As a result there are landscapes linked to leisure (sport and recreation), one of which can be seen in photograph 34. These areas also provide work in service industries.

Core work

32 (a) Study photograph 33. Describe one of the factories and its grounds.
   (b) Compare this photograph with that of New York (photograph 32). Make a list of the differences you can spot between the two working environments. Which would you prefer to work in? Why?

33 (a) Study photograph 34. Describe how the landscape of the coast has been changed to allow the yacht marina to be built.
   (b) Describe the buildings you can see beside the bay. How might they be used?
   (c) In the distance another possible leisure landscape can be seen. What is it?

AMAZONIA: DISCOVERY AND DESTRUCTION

In an earlier part of this chapter we looked at Indian tribes in North America, and saw what happened to them after the arrival of white settlers. In the eastern forests, for instance, almost nothing is left of the original Indian way of life.

Today, in the Amazon Basin in South America, Indian tribes are threatened with the same fate, but this time the changes that are being forced on them are happening much more quickly. At one time, most of the Amazon Basin was covered in a huge thick **tropical rain forest**, which was difficult to explore and broken up only by rivers. Until fairly recently, the Indian tribes which live there had been undisturbed by the modern outside world. Their discovery and the opening up of Amazonia is threatening to destroy the landscape and traditional Indian way of life.

A tribe discovered

This story begins in 1973 when an Amazonian Indian tribe, the Kreen Akrore, were approached

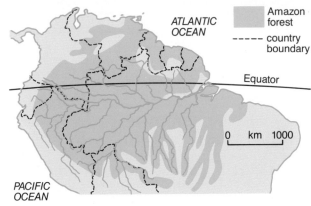

35 *The Amazon Basin*

successfully for the first time. In the following description, the discoverers are flying over the forest in a plane, looking for signs of life.

'The jungle looked blank and meaningless ... then, suddenly, Genario heeled the plane in ... we began to make out a faint crease in the jungle ... There was the village ... a ring of half a dozen huts ... Arrow after arrow rose towards us, ... messages of resistance from the Indian world below... "Walking naked, sleeping on the ground, cutting with stone, roaming the wilderness, they are happier down there than they will be for the next hundred years – even if the contact does go well" said Claudio.'

36 *An Indian settlement in the Amazon forest*

Core work

34 (a) Using your atlas find the area shown on map 35. Which countries does the Amazon flow through?

(b) Measure the approximate size of the Amazon Basin (from east to west and north to south) and compare its size with that of Western Europe and with the Great Plains (see page 42).

35 (a) Photograph 36 shows a small Indian settlement. Study it and describe the scene. Mention such things as the layout of the village.

(b) From what you have learnt about tropical rain forests (Chapter 2), try to list the difficulties of living in such an area.

Extension work

36 Imagine the different feelings of both the discovered Indians and the discoverers. Write a few sentences to describe them.

37 Within a few years of the discovery of the Kreen Akrore, a film was made about them which was watched by millions of television viewers. What good and bad things could result from camera teams and researchers making such a film?

The traditional Indian way of life

As with the North American Indians in the past, the food the Indians eat comes mostly from hunting, fishing, collecting and a little farming. The thick forest offers a wide range of game for the Indians to hunt with their blowpipes and bows and arrows (see photograph 37). Fish too are caught with bows and arrows in the shallower water (photograph 38) or by using a poisonous bark crushed and spread on the water to stun the fish. This makes it easier to gather them up from the river.

One of the main problems facing the Indian is the task of clearing the thick forest for growing crops. Stone axes are not much use when it comes to chopping down trees so, in most cases, fire is used. After a few years, the soil is very poor, so the Indians move on to clear a new patch (see picture 39). This is called **shifting cultivation**.

Although the Indians clear the forest to make way for crops, they do so only in small patches. They do not destroy trees or land on a massive scale, and the abandoned small clearings eventually grow new forest.

38 An Indian fishing with bow and arrow

37 An Indian hunting party

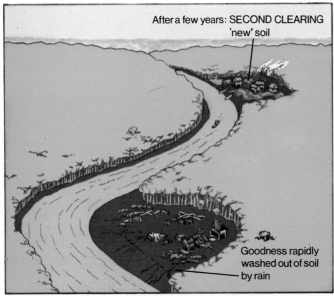

39 Shifting cultivation

Core work

38 (a) Hunting is usually done is groups (see photograph 37). Why do you think this is?
(b) There is very little danger of wild animals in the forest being hunted out altogether. Why?
39 (a) Why do you think the Indian method of farming is called 'shifting cultivation'?
(b) Describe in your own words what is shown in the pictures in 39.

More recent settlers in Amazonia

The Amazon Basin is not only inhabited by Indian tribes. For many decades, settlers have been gradually spreading into the heart of the forest. In the past, settlements tended to be set up beside rivers because these provided the easiest way of moving through the forest. Some people came to exploit the rubber trees in the forest. Poor farmers from other parts of Brazil came because they thought the huge unused region might provide a better living.

In the 1960s, roads began to be built through the thick forest and, since then, many more settlers and foresters have come to exploit its resources. Amazonia is now being opened up for development at a fast rate. Valuable minerals such as tin, iron, manganese and bauxite are being mined. Huge new ranches and plantations are being set up.

Two activities, in particular, drastically change the natural landscape of the Amazon:
(1) the building of the Trans-Amazonian Highway (see map 40) over 5000 km long;
(2) the clearing of huge areas of forest for more and more ranching estates and plantations.

When large areas of forest are cleared (see photograph 41) the heavy tropical rains wash away the plant life and fertile top layer of the soil. This can lead to serious soil erosion and flooding. Also, much of the world's supply of hardwood trees is being destroyed. These are not easily replaced because hardwood trees take 60

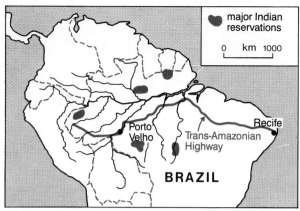

40 The Trans-Amazonian Highway

What hope for these Indians?

Discovery and the development of Amazonia have meant that the Indians have been brought into contact with outsiders, many of whom do not understand or respect the Indians' way of life. As a result, the Indians have suffered poor treatment and many have died from 'Western' diseases caught from new settlers.

In an attempt to protect the Indians from such dangers, the area in which many of them lived was made into a reservation called Xingu Park. The idea behind this was to allow the Indians to carry on their way of life, undisturbed by outside interference. Unfortunately, the plan failed. Roads are being built through the park (see map 42), and settlers are moving in to clear the forest for cattle ranching.

Already, the customs, habits and dress of the Indians are beginning to change, as they learn about the ways of life of the new settlers. (See photographs 43–45.) Many have given up their traditional skills and ways of life. Instead, they beg by the side of the new roads.

This story is similar to the one of the Forest Indians of North America's east coast. In the Amazon, time is fast running out for the traditional Indian way of life.

years to grow to maturity. The loss of great forests might also greatly change the oxygen supply and weather of the world.

Core work

40 (a) List the pull and push reasons for people moving into the Amazon Basin.
   (b) Suggest reasons for the Trans-Amazonian Highway being built through this thinly populated area.
   (c) Why does massive forest clearance cause soil erosion?

Extension work

41 Find out from a library, or Friends of the Earth, how important the Amazon forest is for our oxygen supply and for the future of medical discoveries.

41 A cleared area of forest

42 Change in Xingu Park

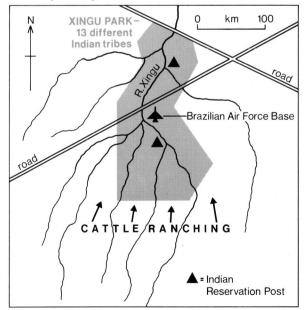

43 An Indian chief discusses a land problem

45 Tourists at an Indian village

44 An Indian woman with tourist crafts

Core work

42 (a) Why are the diseases brought by the settlers so deadly to the Indians?
   (b) Using photographs 43–45 to help you list some of the ways that the Indians' traditional way of life is being changed by recent settlers.

Extension work

43 Do you think it is important to protect the Indians from other ways of life? If so, how would you go about it?

In this chapter we have seen how natural landscapes that were difficult or empty areas, unchanged by people until relatively recently, became inhabited. We saw how these peoples began to change the landscape to make a living from it. These changes have become greater and faster as people have developed more advanced machines and techniques. Developments in mining and factories have added to the changes made by farmers.

In North America we saw Indian lifestyles replaced almost completely by European settlers' ways. Only in areas that were difficult for European farming did traditional lifestyles and landscapes remain. Even in the difficult mountain areas, where farming did not make many changes, mining has resulted in the landscape being changed. The development of cities and industrial areas caused further changing of the natural landscape.

Changes seen over one hundred years in North America are now being seen in the Amazonia area of South America. This time, the speed of change is much greater because of the introduction of technology from more developed areas of the world. This is completely changing the lifestyle of Indians, or causing their tribes to be wiped out. The natural landscape is becoming drastically altered and sometimes totally devastated by the changes.

In the next chapter we shall look at places where the actions of people have caused even greater and more dramatic changes to the landscape in a small area.

4 Landscapes dramatically changed by people

The first two chapters of this book described how the earth's surface was created and then changed by nature's forces. The third chapter described how some of nature's emptier spaces have been developed by settlers, at different times, speeds, and in different ways. This chapter will show how a rural area and an urban area of the world have been dramatically changed by people. The changes have only been possible because of advanced engineering and technical knowledge, and because large sums of money were available to pay for them.

In the Snowy Mountain water scheme in south-east Australia, a massive hydro-electric project allowed water to be moved from wetter areas to drier areas so that farming could be developed. In Hong Kong, people have completely changed the landscape by building a large city on steep barren hillsides.

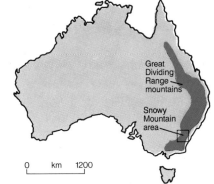

1 The position of the Snow Mountain scheme

THE SNOWY MOUNTAIN SCHEME

The main task of any country is to feed its people. As population increases, so does the need to grow more crops, but not all land is suitable for farming. In part of south-east Australia, the climate is too dry to enable crops to be grown on a large scale, unless extra water can be

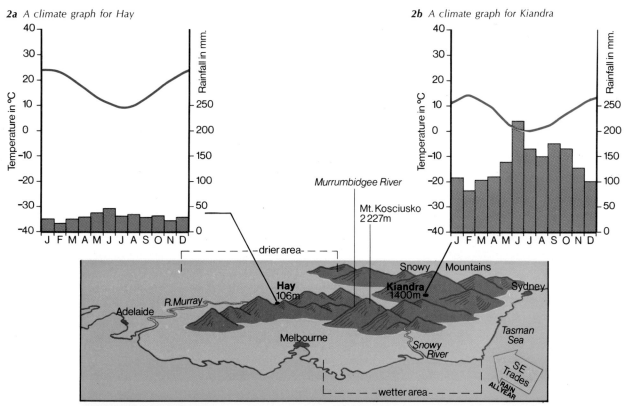

2a A climate graph for Hay

2b A climate graph for Kiandra

2c The Snowy Mountain area

brought in from areas which have more than enough. In recent years, this has been done as a result of the Snowy Mountain scheme. In this section, we shall look at the scheme and the ways in which it has led to great changes in the landscape.

The Snowy Mountain area is shown on map 1 and diagram 2. As the climate graph for Hay (2a) shows, the Murray Basin on the west side of this area does not get enough rainfall to make crop growing possible. It lies in the **rainshadow** (see diagram 3) of the Snowy Mountains. But the mountains themselves, and the area to the east, are much wetter. The water of the Snowy River has therefore been stopped from flowing south by damming, and has been diverted west (see diagram 4). This allows land in the Murray Basin to be irrigated, thereby helping the farmers to get increased yields from the land. Map 5 shows the area which has benefited.

3 *The rainshadow effect (western side) of the Snowy Mountains*

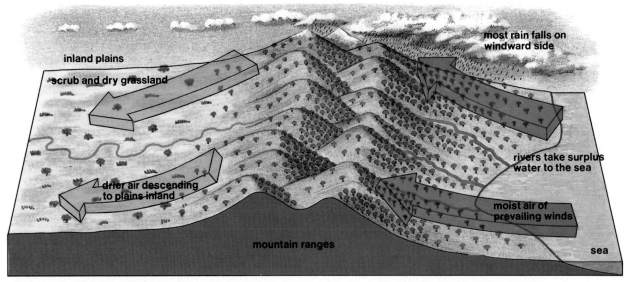

4 *Irrigation and hydro-electric power developments in the Snowy Mountains*

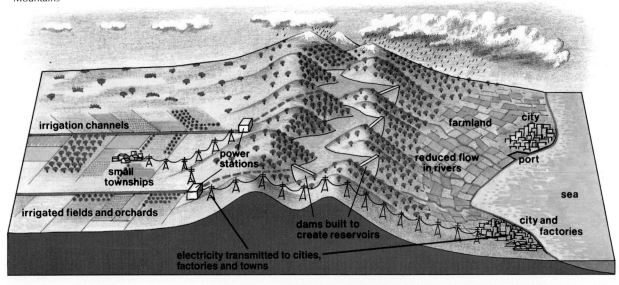

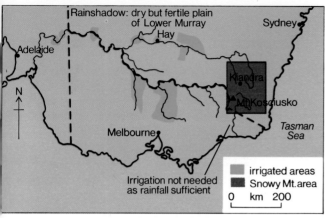

5 *South-east Australia: the area affected by the Snowy Mountain scheme*

Core work

1 Look at map 1, diagram 2 and an atlas.
  (a) Where is the Snowy Mountain Range?
  (b) Which rivers in the area flow west?
  (c) Which river flows south?
2 Study diagram 2.
  (a) What height above sea level is (i) Hay, (ii) Kiandra?
  (b) Using graphs 2a and 2b, work out (roughly) the yearly rainfall for (i) Hay, (ii) Kiandra.
  (c) Describe the difference in height and rainfall in the two places.
  (d) Compare rainfall with temperature on the graphs.
    (i) When does most rainfall occur in Kiandra?
    (ii) When are temperature highest in Kiandra and Hay?
3 Look carefully at diagrams 3 and 4. List the differences between the two.

Extension work

4 Find out and describe how relief rainfall is formed.
5 Explain why the temperature graphs for Hay and Kiandra are very different from the temperature graphs for British towns.

Hydro-electric power

As well as providing water for irrigation, the Snowy Mountains scheme also supplies water to generate much needed hydro-electric power. The Snowy Mountain Hydro-Electric Power Act was passed in 1949, and work began on diverting the waters of the Snowy River inland through tunnel systems into the Murray River, instead of allowing it to flow unhampered into the Tasman Sea. Map 6 shows the completed Snowy Mountain scheme with the dams, the reservoirs formed by them, and the tunnels that divert the water westwards. In its journey through the tunnels, the water falls a total of 790 metres and generates 3.7 million kilowatts of electricity. The water then passes on and is used to irrigate about 800 000 hectares of land.

The total cost of the Snowy Mountain hydro-electric power scheme amounted to about £535

6 *The completed Snowy Mountain scheme*

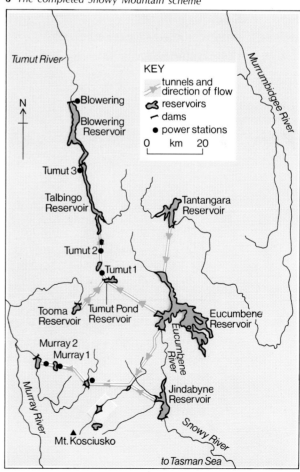

63

million. Revenue from electricity sales meets all costs, including interest, repayment of capital, maintenance and operation of its works.

The electricity produced by the Snowy Mountain scheme is fed into the **power grid** system of the two states of New South Wales and Victoria. This power is used to light thousands of homes, and to drive machines used on farms and in factories and industry.

Core work

6 Study map 6.
 (a) Name the reservoir created by the damming of the Snowy River.
 (b) How is the water taken westward from this reservoir?
 (c) To which river is the water taken?
 (d) Name the largest reservoir in the scheme.
 (e) Name the two sources of water for Eucumbene Reservoir.
7 (a) What is a power grid system?
 (b) What is the power grid system in Britain called?

7 *An underground tunnel being excavated*

8 *Inside Tumut 2 power station*

Extension work

8 Using your atlas, draw an outline map of south-eastern Australia into your workbook. On it, mark and name the eastern cities closest to the Snowy Mountain scheme. These are the cities which will benefit most from the electricity.
9 The electricity produced is for 'peak period' power. Find out what this means.
10 Study Table 1 which shows the scale and importance of the Snowy Mountain scheme. Try to find out about another hydro-electric scheme. (Aswan and Hoover are two examples.) Draw up a table similar to Table 1 and fill in the information for the scheme you chose. Which is the larger and more costly scheme?

Table 1 Some statistics for the Snowy Mountain scheme

Large dams	16
Power stations	7
Largest power station (Tumut 3)	1.5 million kW
Pumping stations	2
Aqueducts	80 km
Tunnels	160 km
Largest tunnel	24.32 km
Cost	£535 million
Total output	3.7 million kW
Time taken to build	25 years

9 Tumut 3 dam and reservoir

Photograph 7 shows an underground tunnel in the scheme, being excavated to carry water. Photograph 8 shows the inside of Tumut 2 power station (built underground). Photograph 9 shows the dam at Tumut 3. Find Tumut 2 and Tumut 3 on map 6.

Core work

11 Look at photograph 7.
   (a) What has the tunnel been carved out of?
   (b) Estimate the width of the tunnel.
   (c) Why do you think it is necessary to have such a broad tunnel?
   (d) Why are tunnels needed to take the water between the rivers and Eucumbene Reservoir?
12 Look at photograph 8.
   (a) Describe the shape of the power station.
   (b) Why is concrete used for the walls of the station?
   (c) The four large machines on the floor of the power station are turbines. What jobs do these machines do?

Extension work

13 Look at photograph 9.
   (a) Think about what this area must have looked like before the dam was built. Try to describe it.
   (b) Make a tracing from the photograph, showing the valley, reservoir, dam, pipes (penstocks). Add notes to describe and explain what can be seen.

Irrigation

Since the Snowy Mountain scheme came into operation, the amount of irrigated land in south-east Australia has more than doubled (see Table 2). In the Murrumbidgee Valley alone, 450 000 hectares are now irrigated.

Table 2 The amount of irrigated land in three states of Australia

State	1957	Present day
Victoria	342 000 ha	695 788 ha
New South Wales	210 000 ha	647 705 ha
South Australia	27 200 ha	74 663 ha

This huge increase in the supply of irrigation water was essential for Australia. Production had to be increased to help the country to support its increasing population without reducing its exports. Farms in areas which are irrigated can produce more fodder crops to feed lambs, beef and dairy cattle. They can also grow crops such as rice as well as increasing the yields of vegetables, fruit trees, vines, oil seeds, tobacco and cotton.

Core work

14 The irrigation water is supplied free to farmers. Why do you think the government does this?

Extension work

15 Why is it important for Australia to export food?

Whitley Glen: an irrigated farm in south-east Australia

One farm that has benefited from the additional irrigation water is Whitley Glen. The farm is situated in the Murrumbidgee Valley, and takes its water from the Murrumbidgee River. It is one of the largest horticultural farms: 21.6 hectares in extent. 18 hectares are planted with over 4000 orange trees. The farm is on fairly level land: there is less than 1 metre difference between the highest and lowest parts of the farm. The soil is mainly well-drained sandy loam.

Irrigation water is brought close to the farm by a main channel like the one shown in photograph 11. The flow of irrigation water to the farm is monitored as it passes over a Dethridge wheel (a meter) from the main channel to the farmer's own channel (see photograph 11). From the farmer's own channel, the water is led into irrigation furrows, made by a plough (see photograph 12).

Irrigation begins in August. In summer the land has to be irrigated every 16 to 18 days, unless it rains heavily enough. The correct amount of water is very important when the fruit is setting, as too much can spoil the crop. It takes three days and two nights to irrigate the whole farm.

11 A main irrigation channel and Dethridge wheel

12 Plough and furrow irrigation at Whitley Glen

13 Spray irrigation

10 An aerial sketch of Whitley Glen

Other farms and orchards in the area produce different crops and use different irrigation methods. Some farmers have installed spray irrigation equipment (photograph 13) which consists of pipes with small rotary nozzles each of which spray a small area. This method of irrigation is expensive to install but has several advantages.

Core work

16 Roughly how many orange trees are grown per hectare on Whitley Glen?
17 Study sketch 10 of Whitley Glen.
   (a) Describe the general layout of the farm.
   (b) What buildings are there?
18 What are some of the difficulties of building a main irrigation channel?
19 What advantages do you think the spray irrigation method would have compared with main channel irrigation?

Extension work

20 (a) Suggest a reason why the flow of water is monitored, even though it is supplied free.
   (b) Describe the plough and furrow method of irrigation shown in photograph 12.

Graph 14 shows the climate graph for Whitley Glen. Oranges need 750–1000 mm water (from precipitation or irrigation) per year. This allows for water lost through evaporation, transpiration (through leaves of plant, like sweating) and seepage into the soil.

Core work

21 Study graph 14.
   (a) Work out (roughly) the total yearly rainfall at Whitley Glen.
   (b) Explain why irrigating the farm is worth while.
   (c) In which months will irrigation be particularly helpful?

Extension work

22 (a) How much irrigation water will be required each year?
   (b) Will the kind of soil have any effect on the need for water?

Water for towns and industry

There is a third benefit from the Snowy Mountain scheme. Water is also used for towns and industry, and it is hoped that, in the future, more small townships like Mildura (photograph 15) will develop in the Murray Basin. With increased power and more local farm produce, many of these settlements may become industrial centres. They would also become more important as social centres.

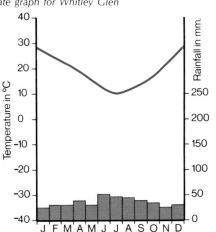

14 A climate graph for Whitley Glen

15 The township of Mildura, Murray Basin area

Core work

23 (a) How will the development of townships slow down the movement of farmers' families to the great eastern cities?
(b) How will this benefit the area generally?
(c) Study photograph 15 and describe the township shown.

Summarising the Snowy Mountain scheme

Core work

24 Copy diagram 16 into your book. Make a summary of what you know about the Snowy Mountain scheme by writing short notes in each box. Diagrams 3 and 4 will help you.
   This scheme is a 'system'. Things feed into it, are changed or used, and other things are produced out of it. It is an input–output system.

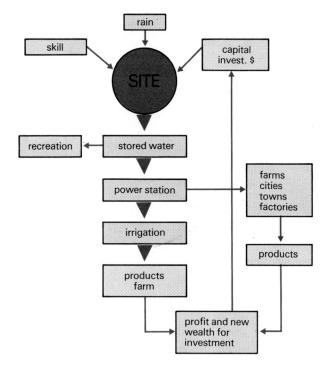

17 A flow diagram to show the inputs and outputs of the Snowy Mountain scheme

25 Compare diagram 17 with your summary diagram. What does it include that yours does not show?

16 Summary cross-section diagram of the Snowy Mountain area

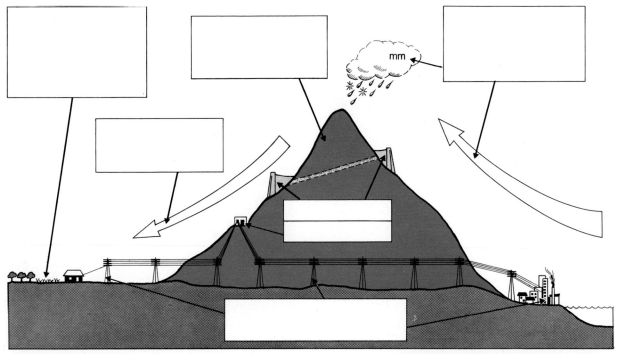

Core work

Study photographs 23–25.

31 (a) Why do you think public transport is important?

(b) Why are ferries needed? (Hint: look back at map 18.)

32 Describe what Hong Kong is like today.

Extension work

33 Compare Hong Kong with the place where you live. Use these headings: Roads, Buildings, Signs, Traffic, People, Land.

24 Flats, Aberdeen, Hong Kong

23 Trams in Hong Kong

25 City centre street, Kowloon, Hong Kong

Housing the population

The shape and slopes of the Hong Kong territory have made it very difficult for housing authorities to cope with the dramatic increases in population shown in graph 26. Hong Kong is now one of the world's most densely populated areas.

During the years after 1945, Chinese immigrants flooded into Hong Kong. To begin with, food and relief centres were set up throughout Hong Kong by the government and voluntary organisations. Tenement accommodation was partitioned to squeeze in extra people, and make-shift dwellings appeared on roofs to take the overflow.

Temporary dwellings made out of odd pieces of corrugated iron, wood and cardboard, called **shanty dwellings**, spread over the hillsides (photograph 27). Many families set up home on boats in the harbour (photograph 28). About 100 000 people still live as squatters in these conditions.

The uncontrolled building of shanty settlements brought fires and outbreaks of disease. To solve these problems, the Hong Kong Government started to build cheap blocks of flats in 1953. Production of housing continues today at a rate of 35 000 flats a year. Shortage of suitable land has made it necessary to build most of the new housing on reclaimed land. Land use is maximised by building skyscraper blocks of up to 36 storeys (see photograph 29).

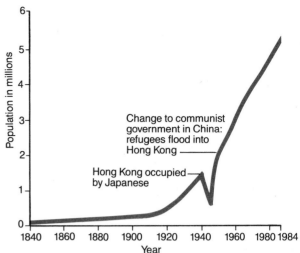

26 *A population graph for Hong Kong*

27 *Shanty dwellings where many new immigrants live*

28 *Sampans: Hong Kong boat houses*

29 *Block of flats*

Core work

34 (a) What do you think caused the sharp drop in Hong Kong's population in the early 1940s?
   (b) How many people left during that time?
35 (a) List the ways in which the new settlers were accommodated in Hong Kong after 1945.
   (b) Describe the types of houses shown in photographs 27 and 28.
36 (a) Describe the type of housing (shown in photograph 29) built to replace the make-shift dwellings.
   (b) What are the advantages of building multi-storey flats in an area like Hong Kong?

Extension work

37 (a) Why was most of the new housing built on reclaimed land?
   (b) Photographs 27–29 all show 'high-density housing'. What do you think this means?
   (c) Would you like to live in these conditions? Give reasons for your answer.

30 Sha Tin New Town

To cope with Hong Kong's enormous population, a housing programme was set up in 1972 with the aim of providing housing for 1.8 million people. Most were to be housed in three New Towns in the New Territories north of Kowloon (see map 18): Tsuen Wan, Tuen Mun and Sha Tin (photograph 30). Since 1972 the programme has been extended considerably. There are now eight New Towns, all built on reclaimed land in the New Territories. In 1985 they housed a total population of 1.7 million which is expected to grow to 2.5 million by 1991.

Each of the New Towns is planned to provide a full range of shopping facilities, community centres, schools, clinics, etc. Land for industry is also provided so that residents can work locally.

Core work

38 Study photograph 30 of Sha Tin.
   (a) Describe the land on which the New Town has been built.
   (b) How does it differ from the building land on Hong Kong Island (see photograph 22)?
   (c) Describe the type of housing that has been built.
   (d) What do you notice about the main roads compared with those in the main city in Hong Kong?
39 Copy the diagram below. Add labels to explain what is happening.

Trade and industry

When immigrants began to arrive in Hong Kong after 1945, there was little work available for them. But they could be trained to do skilled work. This large workforce was an important resource which was used by wealthy companies and factory owners. It enabled Hong Kong to become an important centre for manufacturing industry.

The textile and clothing industry was one of the first, and it is still Hong Kong's most important manufacturing industry. 42% of the total industrial workforce are employed in it, and in 1984 it accounted for about 40% (by value) of Hong Kong's total exports. A wide range of yarns, fabrics and clothing is produced.

Among the other early manufacturing industries were footwear, household utensils, electric torches and toys. Gradually new industries were introduced such as plastics, electrical goods and electronics. The electronics industry has expanded rapidly in recent years. In 1984 there were 1342 factories employing about 100 000 workers, and manufacturing radios, cassette recorders, Hi Fi systems, television sets, microcomputers, etc. In the early days of industrial development, Hong Kong imitated western products, but it now produces many original and high-quality goods.

The tremendous industrial development that took place in Hong Kong can be seen from the statistics in Table 1.

31 A multi-storey factory building

Table 1 Industry statistics, Hong Kong

	1962	1972	1976	1984
Registered factories	7 305	21 386	35 760	48 038
Industrial workers	297 897	619 684	778 778	898 947
Domestic imports (million Hong Kong $)*	3 317	15 245	32 629	137 936

*HK$10 million is approximately £1 million

32 A large factory

In some of the urban areas of Hong Kong there are over 250 factories per square kilometre. Many of these are quite small and are in multi-storey factory buildings (see photograph 31). Others are very large, like the one shown in photograph 32, and are built away from the urban area.

Table 2 Employment in Hong Kong, 1984

Industry	Number of workers
Manufacturing	900 000
Construction	68 000
Wholesale/retail restaurants/hotels	556 000
Finance and business	168 000
Government services	172 000
Fishing and farming	28 000

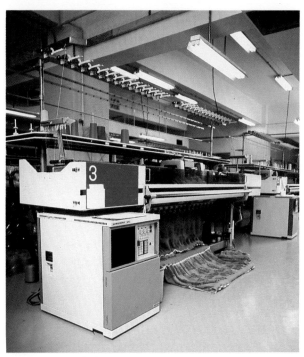

33 Inside a textile factory

34 Inside an electronics factory

Core work

40 Study Table 1.
  (a) What do the statistics show?
  (b) Why do you think industrial output has continued to grow so rapidly in Hong Kong?
41 (a) Why is the kind of factory building shown in photograph 31 well suited to crowded urban areas?
  (b) What is the main thing produced by the factories in the picture?
42 Study photograph 33. Describe the scene.
43 Study photograph 34.
  (a) Why do you think the electronics factory needs a large number of workers?
  (b) What kind of work is being done by the women?
44 Study Table 2.
  (a) List the industries in order of importance.
  (b) How do most people earn their living in Hong Kong?
  (c) Why do fishing and farming employ so few?

Extension work

45 (a) Draw a bar graph to show the employment figures in Table 2.
  (b) Obtain similar figures for your country or region. Draw a bar graph for these.
  (c) Compare the two graphs and try to explain similarities and differences.
46 Look at photograph 32.
  (a) What are the advantages of building large factories outside the urban area?
  (b) Describe the land shown in the picture.
  (c) Why is the factory built on reclaimed land?

Most of Hong Kong's manufactured goods are exported through the port. Each year more than 11 800 ocean-going ships enter and leave the port, loading and unloading about 48 million tonnes of cargo.

The well-equipped ocean terminal (photograph 35) enables a very fast turnaround of ships in the port. Four medium-sized ships can be handled at one time and, on average, 140

35 Ocean terminal

ocean-going ships are in the port daily. The modern container terminal (photograph 36) takes up to six ocean-going ships at one time, reducing the time taken for loading and unloading.

Hong Kong International Airport is one of the busiest in the world. It is linked to all parts of the world by 31 international airlines, which handle more than 54 000 scheduled aircraft movements a year. (This compares with 85 000 aircraft movements handled at Scotland's twelve main airports in a year, and 278 000 at Heathrow Airport near London.) In addition, the airport has many non-scheduled passenger and cargo charter flights.

The runway (photograph 37), built on reclaimed land, extends 3390 metres into the harbour (see map 18). The airport is less than 5 km from the centre of Kowloon.

Core work

47 Study photograph 35.
   (a) Describe the scene at the ocean terminal.
   (b) Name two ways in which the ships at the ocean terminal are unloading their cargo.
   (c) Why is it important for Hong Kong to be able to handle ships quickly?
48 Look at photograph 37.
   (a) Why was the runway built out into the sea?
   (b) What are the disadvantages of locating a major airport so close to a large urban area?
   (c) What are the advantages?

Extension work

49 Choose a city centre with a commercial area near your home, or an industrial region in Europe, and find out how growth and development have changed the landscape over the years.
Write up a report in the form of a newspaper article.

36 Container terminal

37 Airport runway on reclaimed land

Most of the land area of Hong Kong (the New Territories: the area north of Kowloon and the islands) was leased to Britain from China in 1898 for 99 years. The lease is therefore due to run out in 1997. In the early 1980s there was concern over what would happen to Hong Kong when the lease ran out. In September 1982, Britain and China agreed to enter into talks on the future of Hong Kong, with the aim of maintaining Hong Kong's stability and economic success.

After two years of negotiations, an agreement was signed by Britain and China in December 1984. This agreement allows Hong Kong to keep its present economic, legal and social systems and way of life for 50 years after 1997. Hong Kong will become a Special Administrative Region of China, but will continue to have a capitalist system rather than the communist system of mainland China.

Few other areas of similar size and population can match Hong Kong's success in international trade. Nor can they illustrate better the way people can change a difficult living environment into a bustling industrial and commercial city.

Core work

50 Why do you think it was important for the British and Chinese governments to reach an agreement so early before the lease runs out?
51 Imagine you owned a factory in Hong Kong in 1985 and wanted to build a larger one. How would your decisions have been affected
   (a) if no agreement on the future of Hong Kong had been reached?
   (b) now that the present agreement has been reached?
52 Why do you think the communist government of China is likely to allow the capitalist system in Hong Kong to continue after 1997?

5 People: patterns and problems

1 A Victorian family

2 A present-day family

BRITAIN'S POPULATION IN THE PAST AND PRESENT

'Population' usually means the number of people who live in a place at any particular time. It can be affected by the way people live, the conditions they live in, and the type of land they live on. The family in photograph 1 lived at the time of Queen Victoria. The family in photograph 2 looks very different. The clothes and hairstyles have changed, but the biggest difference is in the size of the family. Families of eight or nine children were very common in Victorian times. Today, families are usually much smaller.

Core work

1. (a) Look at photographs 1 and 2. How many people are there in each family? Can you guess who are the parents and who are the children? How do you know?
   (b) What could be the good things about living in a large family in the past?
   (c) What could be the bad things about it?
   (d) What could be the good things about living in a large family today?
   (e) What could be the bad things about it?
2. (a) Write down the problems which would face a poor couple bringing up a large family in the past.
   (b) How would it have been easier for a rich family to raise children in the past?
3. (a) Write down the different ways children today cost their parents money.
   (b) Write down how parents today manage to raise children without the help of servants.

Extension work

4. (a) If each child in the families shown in (i) photograph 1 and (ii) photograph 2 had as many children as their parents had, how many grandchildren would there be?
   (b) What would be the difference between the two families' sizes after two generations if the same happened again?
5. (a) Ask five people in your class how many children are in their family. Draw a graph of your findings.

(b) Ask the same five people to find out how many children were in their parents' families. Draw a graph of your findings.
(c) Comment on the pattern the graphs show.
6 Find out about the size of families in your area in the past. You could ask old people, or look at churchyards or burial grounds, or records in your local library. Record your findings.

In the past, life could be dangerous and short. Let us look at the fate of two families who lived about 300 years ago in Colyton, a small market town in Devon (see map 3).

William Hore, a shoemaker, and his wife Johane had eleven children in the twenty years of their marriage. Two died as babies; another girl, Charyttye, died young; and the rest all lived to leave the village as adults, except Eddythe who remained in Colyton. Solomon Bird was a weaver, thirty years younger than William Hore. His family was not so fortunate. Two children, Marie and Robert, died young, and in one fortnight in the summer of 1646 plague killed Solomon, his wife and their three remaining children. Only one son, Rawlynn, lived. He had left Colyton five years earlier to get married.

3 The position of Colyton

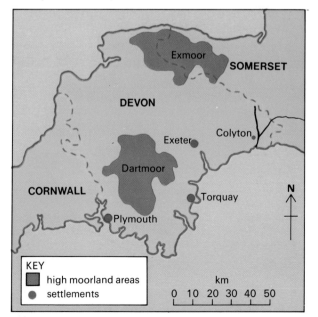

Core work

7 (a) How many children were born in the Hore family?
(b) How many children in that family died?
(c) How many children in that family stayed in Colyton as adults?
8 (a) How many children did Solomon Bird have?
(b) How did he, his wife and his children die?
(c) Who in his family lived? Why did he not die?

Extension work

9 (a) Write a story of the death of the Bird family as if you were writing a report for a newspaper of the time.
(b) What reasons do you think there might have been for children to die in the past?

Birth rate and death rate

The term **birth rate** means the number of children born each year. In the past the birth rate in Britain was higher than it is today. Families were larger for several reasons. Many people were farmers who wanted sons to help them work the land. Many were afraid of losing their children in the 'plagues' which regularly swept the country, killing thousands of people. Many hoped that their children would live to care for them when they were old, since there were no pensions or homes for old people. Many may not have wanted to have so many children but had them because they did not know how to prevent children from being born.

The **death rate** is the term given to the number of people who die each year. In the past, it was affected by disease and famine. The birth rate and the death rate were often linked if there was a shortage of food when too many children were born. Hunger would kill the weakest, or disease would sweep the village, killing old and young

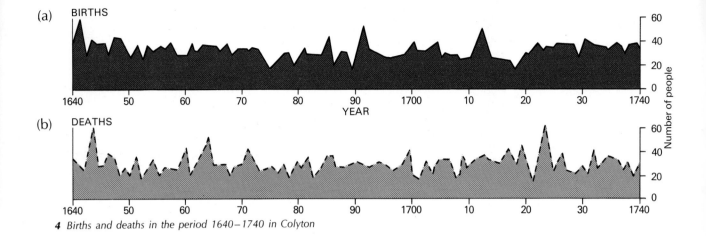

4 Births and deaths in the period 1640–1740 in Colyton

alike. Many people were dead by the age of forty: in other words, their **life expectancy** was low.

Graphs 4a and b show that, in the past, the number of people being born was high and the number of people dying each year could vary very greatly.

Core work

10 Write a sentence in your own words to explain what (a) birth rate, (b) death rate and (c) life expectancy each mean.
11 (a) Trace the patterns in graphs 4a and b with your finger. What do you notice about the relation between the two?
   (b) Look at graph 4a. Name three years when births were very high, and three when they were very low.
   (c) Look at graph 4b. Name three years when deaths were very high and three when they were very low.
   (d) Name a five-year period when births were stable (stayed fairly level).
   (e) Name a five-year period when deaths were stable.
12 Explain in your own words why births and deaths varied so much 300 years ago.

The population begins to grow

After 1800 there was a rapid increase in the population of Britain. This great change happened because improvements in medicine, food and living conditions meant that fewer people were dying. People were living longer than before. There was no change in the birth rate. It continued to be high, and families were large.

Core work

13 (a) Write a sentence saying what happened to Britain's population after 1800.
   (b) Make a tracing of graph 5. Put the following labels at the correct places on your tracing.
   Population low and stable
   Population rising very fast
   Population rising slowly
   Population high and stable

5 Population in Britain

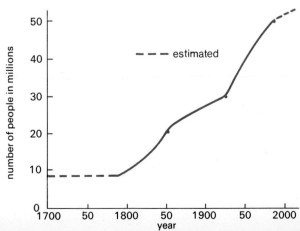

Extension work

14 (a) Graph 5 shows that the population in Britain before 1800 did not change greatly. What does this tell you about the birth rate and death rate at that time?
   (b) For population growth to take place after 1800 there had to be a change in the birth rate or the death rate. What would the change have to be? Explain your answer.

Core work

15 (a) How long might someone in Britain hope to live in 1840?
   (b) How long might someone in Britain hope to live in 1980?
   (c) Using information from diagram 7, write three short paragraphs headed Medical, Food and Social to explain why people began to live longer after 1800.
   (d) Why is it that when people live longer the population gets bigger?

16 (a) For each bar in graph 6 use the scale to help you write down how long people could expect to live at that time.
   (b) What do you notice about the pattern of the graph as you move towards the present day?

Extension work

17 (a) Make two lists, one of the things **families** provide children with as they grow, and the other of things provided by the **government** for children.
   (b) Discuss in a small group, or your class, what people have included in their lists. Which things did most people include? Was anything mentioned you had not thought of?

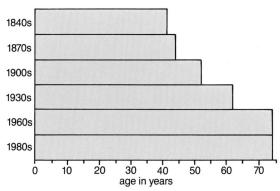

6 Life expectancy in Britain after 1840

As graph 6 shows, life expectancy continued to increase from 1840 onwards. Diagram 7 illustrates the main reasons for this.

7 Why did people begin to live longer?

- new hospitals and medical school
- better hygiene
- vaccination
- new treatments
- more food from abroad
- improved food storage
- improved farming produces more food
- **MEDICAL**
- **FOOD**
- **SOCIAL**
- free health care
- demolition of unfit housing
- improved sanitation and water supply
- better pay and education for all

Understanding population pyramids

A population pyramid is a bar graph which shows how a country's population is made up in terms of age and sex. There are separate graphs for males and females, each showing the numbers of people in different age groups (0–9 years, 10–19 years, etc.) Graphs 8a and b show how the population of Britain was made up in 1901. At this time the population was increasing fast because the birth rate was high and the death rate was getting lower.

8 Graphs to show the age and number of (a) males in 1901, and (b) females in 1901

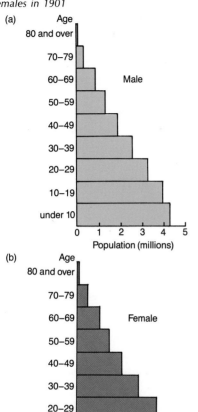

Core work

18 Look carefully at graphs 8a and b.
  (a) Which age group was largest for (i) males and (ii) females in Britain in 1901?
  (b) Which age group was smallest for (i) males and (ii) females?
  (c) Describe the pattern which each of the graphs shows.

Extension work

19 (a) Make a tracing of graph 8a. Label it MALE.
  (b) Make a tracing of graph 8b. Label it FEMALE.
  (c) Fit both graphs over graph 9. Do they match? How can you make the male graph fit?
  (d) Mark in colour on your tracings the line along which the graphs have been joined.
  (e) Explain why graph 9 is widest at the base and narrowest at the top.
  (f) Look carefully at graph 9. Which sex tended to live longer? How can you tell?

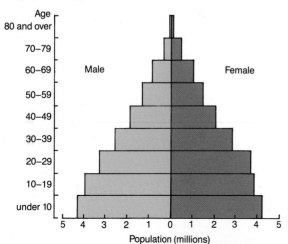

9 A population pyramid for 1901

Graph 10 shows the population pyramid for Britain in 1980. The shape of this pyramid is very different from the one in graph 9. This is because families are now much smaller than before, and people live longer. Britain has been wealthy enough to look after its population, and great advances have been made in curing and caring for people of all ages.

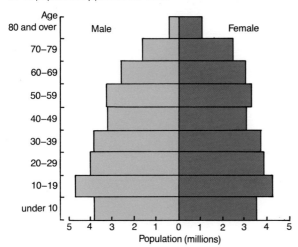

10 A population pyramid for 1980

Providing for Britain's old people

A new problem facing Britain and other rich nations of the world is that old people now make up a larger proportion of the population than before. Graphs 9 and 10 showed this.

Old people have special needs. They have greater difficulty in moving around than younger people do. They may need help to manage their homes; some may not be able to look after themselves properly. Winter may affect them badly and their health may be more delicate. On the other hand, they may have more time to devote to their interests and to other people.

In a country such as Britain, where there are greater and greater numbers of old people, extra services must be provided. Old people in Britain receive government pensions. They usually need more hospital places, medicines and help from social and medical workers. Sometimes it is also necessary to set up social centres, which are run and equipped by government money, so that old people who are lonely, cold or bored have a place to go.

It is not difficult for the government to meet these needs, as long as most of the population are working and paying taxes which will pay for the necessary services. The more old people there are, however, the more services are required, and the smaller the proportion of the population is which works to pay for them. This might mean that a country could not afford to care properly for its old people.

Core work

20 (a) Trace the side and base scales for graph 9 and draw in the centre line (at 0 on the base scale). Now trace the outlines of graphs 9 and 10 onto this base, making sure that the base scale and centre line are correctly in position. This combined graph will show up the differences between the two shapes very clearly. List the differences that you can see.
(b) How can you tell from the graphs that people lived longer in 1980 than in 1901?
(c) What can you say about the birth rate in 1980 compared with 1901?

Extension work

21 The table below shows how family size has changed. Write down some possible reasons for this change.

The number of children in an average British family

1870	1880	1910	1940	1970	1980
6.0	5.4	2.8	2.0	2.2	2.2

Extension work

22 (a) Interview a few old people that you know. Try to find out how they spend the day, and the problems they face in everyday life. Record these findings in your workbook. You should not use names of the people you interview.
(b) How are their lives different from your own or your parents' lives?

Many older people, particularly those who have worked and lived in a large city, decide to move to a new area when they retire. These

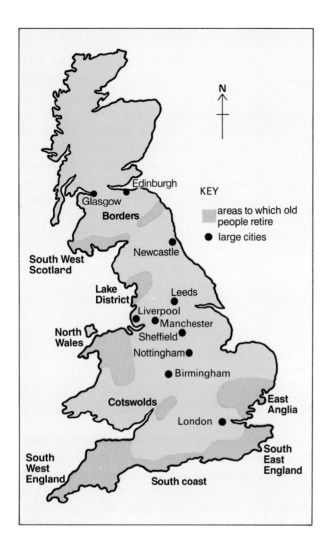

12 Old people in Bournemouth

people usually choose an attractive quiet place in the countryside or on the coast. Map 11 shows the main areas that people retire to in Britain. One of the most popular places is Bournemouth on the south coast of England (see photograph 12).

Core work

23 (a) List the names of the nine retirement areas shown on map 11 in your workbook. Beside each one write the name of a large city which is near it.
(b) Which area is far from any large city?
(c) Look at photograph 12. Write down four reasons why Bournemouth might be attractive to old people.

Extension work

24 What help do old people need which younger people do not?
(a) From their families. (b) From the government.
25 (a) What do you think are the biggest problems old people face in their daily lives? (The information in your answers to Extension work 22(a) should help you with this.)
(b) What special services might need to be provided in areas where many old people choose to live?

IRELAND: FALLING POPULATION, 1840–1970

Normally, the population of most countries is rising, often quite rapidly. But there are times when the population of a country may decrease. This happened in Ireland between about 1840 and 1970, as you can see from graph 13.

Map 14 tells more of the story. Large areas of Ireland lost a fifth of their population between 1846 and 1851, the period when graph 14 begins to dip steeply. What catastrophe happened to empty the land so quickly of its people?

The following description of Ireland at that time will give you clues.

'I had heard of the innumerable shades of green and the darling little thatched cabins ... But nothing I had heard had prepared me for the poverty, the filth, the beggars, the mud huts and that ruined blasted countryside which looked like it had been ruined by some catastrophic war ...
The smell reached us a moment later ... We were looking at the living dead, scarecrows who might once have been men and woman ... There were small scarecrows too with grotesque swollen stomachs and one woman carrying a dead baby with a blackened tongue.'

From *Cashelmara*, by Susan Howatch

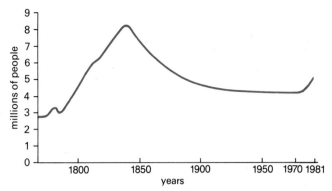

13 Changing population in Ireland

The writer is describing the suffering of people who died from the potato famines of the 1840s when disease killed the potatoes which were the basic food of Ireland, or who died from the typhus epidemics which accompanied the famine. The population had grown in the early 1800s when the introduction of the potato into Ireland provided more and better food. The terrible famines and fever had killed 800 000 people by 1850.

To add to this problem, new ideas about farming were reaching Ireland at the same time. Landowners increased rents, and cleared tenants off the land to allow their own animals to graze. The tenants had nowhere to go. They often faced a grim alternative: to die on the streets of Dublin, or **emigrate** to another country. As Table 1 shows, many chose to leave the country.

Many of the emigrants moved to large cities on the mainland, such as Glasgow and Liverpool. Others sailed to the Americas on the 'coffin' ships, so called because huge numbers of emigrants died in the unhealthy conditions on board. Those who did survive could find land, or could work in the new American cities.

Table 1 Number of Irish emigrants

1780–1845	1848–1855	1914–1918
1 700 000	2 000 000	3 500 000

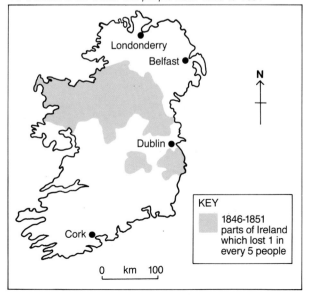

14 Ireland: areas which lost people from 1846 to 1851

Core work

26 (a) What two causes were there for the deaths of so many Irish people in the 1840s?
 (b) Describe the changes in farming which also led to a fall in population.
 (c) Using map 14 and an atlas map, find the names of the areas worst affected by population loss in 1846–1851.

Extension work

27 (a) Use graph 13 to answer the following. What was the approximate population in Ireland in 1800, 1850, 1900 and 1950?
 (b) Look at the graph and say during which century the greatest change in the population took place.

People in Ireland today

Map 15 shows the main towns in Ireland today. An atlas will give you more details. If you compare Ireland with most other flat areas of similar size in Western Europe you will see that Ireland has far fewer big towns in it. Photograph 16 is a scene typical of some parts of **rural** Ireland. Tourists travel far to visit wild 'empty' areas such as these, but, for people who live there permanently, the **isolation** can create problems. Having a tooth filled, catching a train, buying clothes or visiting a hospital can all mean very long journeys to distant towns. There are not enough people in the countryside to make it worthwhile providing **services** there. Jobs are also harder to come by than they are in the towns.

As a result many young people in Ireland leave the countryside to go to the towns (sometimes even other countries). They go in search of a job, and to be within easy reach of all the services that city people take for granted. There are many other parts of the British Isles which show a similar pattern of young people leaving for the cities: for example, the Highlands of Scotland, the Dales of Yorkshire, the Cumbrian Lake District, Exmoor and Dartmoor. All these areas are wild rural landscapes with small populations.

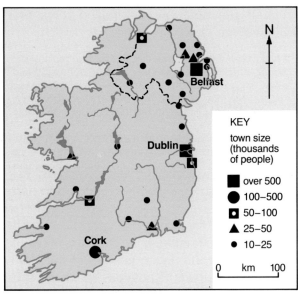

15 Towns and cities in Ireland today

16 Rural Ireland

Core work

28 (a) Look at photograph 16. Write down why a tourist from a city might find this area an attractive place to visit.
   (b) Why might a local young person wish to leave this area?
   (c) What problems would farmers face in such an area?
   (d) What other jobs might there be there?

Extension work

29 Try to find out about an 'empty' area near your own home. Find out how far apart the settlements are, using a map. Try to find out why the area is so empty. Write a report on the area, using drawings or maps, and mention the problems and benefits of 'emptiness'.

30 (a) Draw a diagram like diagram 17. Use map 15 and an atlas to help you complete it.
   (b) What do you notice about the pattern of the number of towns and their sizes?

17

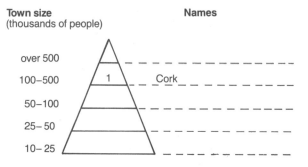

JAPAN: A CROWDED INDUSTRIAL COUNTRY

Japan is similar in area to Britain, but it has 120 million people: twice as many as in Britain. Japan has, on average, nearly 300 people per square kilometre. It is a land of contrasts. Map 18 shows that Japan is a mountainous country, with small pockets of flat land round the coast. Great cities fill much of the flat land (see map 19). Some have grown so large that they have joined together to form **conurbations**. Nearby mountains are steep and difficult to reach and use, so fewer people live there. Those who do tend to live in the valleys or on the lower slopes, and these areas are quite crowded.

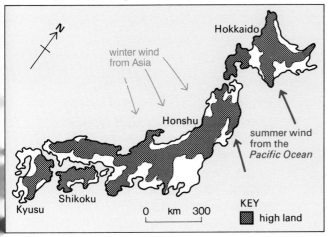

18 Japan: islands, highlands and winds

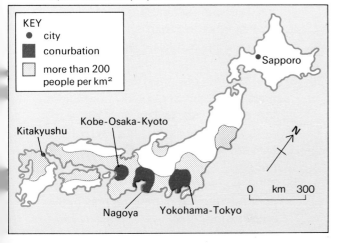

19 Japan: where the people live

Supporting the population

Britain's population grew because the country had enough food to feed its people, and had many **resources** (such as coal, iron ore) to build up industry, and so provide jobs and goods for its people. Japan is not especially rich in resources, and the mountainous land makes farming difficult. But the Japanese people have made the land support many people by very clever use of their resources, and by importing from abroad those they do not have themselves (for example, oil and coal).

The growth of industry in Japan came later than in Britain. It began in the 1800s but was largely destroyed in the Second World War. Rapid growth did not begin until 1960. Before that, most people were living off the land, and were making very little money. Major change began when the Japanese government drew up a plan to double national income. This involved building roads, ports and factories and selling more goods abroad. Japan hoped to reach a total income of roughly £30 000 million by 1970, but instead reached £80 000 million. By 1982 it was £700 000 million.

Core work

31 Write down in a list the names of the four main islands which make up Japan. Choose which of the descriptions in the box below fit which island, and write them beside the correct name in your list.

> The biggest island; the most northerly island; the most mountainous island; the smallest island; the most southerly island; the island with most conurbations; the island where Sapporo is the main city; the island with most people; the island with no main city.

32 (a) Japan does not have many **natural resources**. Find out what this phrase means. Write down your answer.
    (b) How does Japan get resources for industry if it does not have them in its own country?

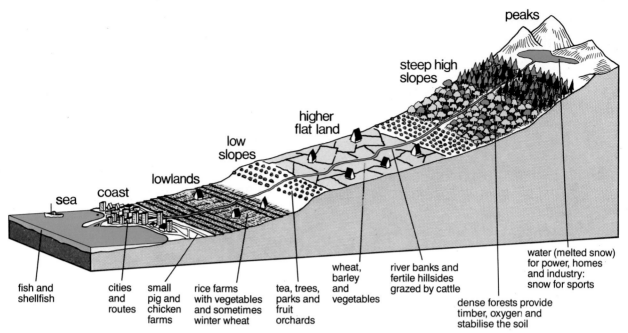

20 The way land is used in Japan

Using every inch of land

In Britain, cities are also usually on the plains and coasts, and farms fill the lowlands. There are great stretches of land, however, where few people live and little use appears to be made of the land. This is not so in Japan. Diagram 20 shows how land is used from the coast to the mountain peaks.

In the past most of Japan's people worked on small farms and grew the food they ate. Today things have changed. Rice and wheat can be imported cheaply. Farmers with small farms can make more profit growing vegetables or rearing chickens and pigs on battery farms. More cattle are kept now that living standards have risen and people can afford more meat in their diet. Small farmers can often make more money working in nearby cities, however. Young people are attracted by the easier working life and facilities the cities offer. Some leave the countryside to work in the towns, returning to the villages for holidays and eventually to retire. Some commute to the towns each day to work.

Core work

33 (a) Copy diagram 20 in your workbook.
(b) Write down three headings:
Flat land, Water, Sloping land.
Under each of these, list the various ways these are used in Japan, using diagram 20 as a guide.

Extension work

34 (a) Try to explain the different ways that flat and sloping land are used at different heights in Japan.
(b) Which land would be most valuable? Why?

35 (a) Try to find out from television programmes or library books about life in the countryside in Japan. Write a report about the work, homes, village life and problems of these country people.
(b) Why do young people in Japan not continue the life of their parents in the countryside?

Coastal industry

Rainfall over the mountains is used to create power in Japan in the same way as in the Snowy Mountain area of Australia (see page 62). Reservoirs in the mountains collect water which is used to turn turbines as it rushes down the mountain sides, to make cheap electricity. These **hydro-electric power stations** tend to be small, however, too small to meet the enormous demand for power from Japanese homes and factories. **Thermal power stations** use a fuel such as coal or oil to make steam to turn the turbines. Map 21 shows where these are found in Japan. They are mainly on or near the coast.

Japan's wealth and ability to support a large population result from its high level of industrialisation. All the conurbations (photograph 22) have major industrial zones which produce all kinds of manufactured goods, from ships and cars to high-technology products such as electronics, computers, chemicals and engineering equipment. Not all firms are very large, but there are many giant industries in Japan. The factories are often located by the coast, sometimes on land reclaimed from the sea (photograph 23).

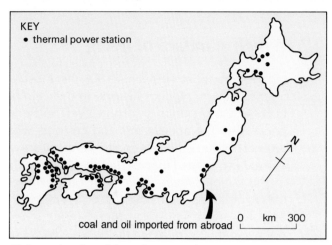

21 Japan: location of thermal power stations

Core work

36 (a) What fuels are used to generate electricity in Japan?
 (b) Why is it sensible to build power stations on the coast, in that case?
 (c) Look back at maps 18 and 19 and suggest two more reasons why the thermal power stations are located where they are.

Extension work

37 Using information from television programmes or library books find out about one modern industry in Japan. Write down how it works and in what ways it is similar to or different from a similar industry in Britain.

22 Industry in a conurbation

23 An industrial site on reclaimed land by the coast

A high standard of living

Japanese families today have for the most part adopted a very modern lifestyle in their clothes, homes, cars, public transport and household goods. But the ancient arts and customs which are part of Japan's history and heritage are still carried on. Some Japanese families, particularly in the country areas, have held on to more of these traditions than others. All Japanese families, however, are able to use modern technology and labour-saving machines which make life much easier than in the past (see photograph 24). Most are also much richer than in the past and, as a result, have a wider choice of how to use their money. Homes are comfortable and well equipped. Successful industry has created the wealth which pays for this.

Japanese cities have fine shopping areas with a vast range of goods for sale (see photograph 25). Traditional foods such as rice, vegetables, fish and pickles are being added to by more meat — hamburgers are especially popular among young people!

24 A Japanese family at home

25 A shopping street in Tokyo

Core work

38 (a) What do you think a visiting Japanese child would find strange or familiar on a visit to your home or town? Write down your ideas.

(b) Use photographs 24 and 25, as well as information you might have from television programmes or library books, to describe an imaginary visit to a Japanese home and town. Remember to include things which you would recognise as similar to those you might have at home, as well as all the things which would be strange.

Extension work

39 What do you think would happen to living standards in Japan if people in other countries stopped buying Japanese products such as cars, televisions or ships? Why? Discuss your answer with your teacher or group.

Problems with success?

As we have seen, Japan is a small and very crowded country which depends on importing food and raw materials from other countries to support its population and industry. In turn it produces and sells goods to other countries. Successful trade and industry has made Japan a very wealthy country. The Japanese want to keep raising their high living standards. To do this they must continue to produce goods which other countries want to buy. This means they must build more factories, offices, homes and shopping facilities.

There is a price to pay for Japan's success, however. The average family house is no more than 46 square metres in area, and many city flats have only 14 square metres of floor space. Measure an area of this size on your classroom floor and compare it with your own home. Small houses like the ones shown in photograph 26 might cost four times as much as a small house in Britain. Since most Japanese homes have no gardens or only a very small garden, at holidays and weekends, parks are crowded with people (photograph 27).

The huge size of the cities means that many people have a journey of up to two hours to get to work each day. Also, the cities suffer badly from air and water pollution, caused mainly by the large numbers of motor vehicles and factories. Efforts are being made to control this, but the pollution continues.

27 A Japanese park

26 A Japanese house

Core work

40 (a) Which of the problems mentioned in the text are caused by
   (i) the number of people in Japan?
   (ii) the growing wealth of the people in Japan?
(b) Which of the problems do you think might be solved? How might they be solved?

Extension work

41 (a) How would measures taken to control pollution affect industry?
(b) How might this in turn affect the living standards of the people?

42 Is the price Japan has paid for wealth too high? Give reasons for your answer.

43 What do you think the future holds for Japan? Write an account of an imaginary visit to Japan which takes place in 50 years' time. What changes would you expect to see? You might imagine, for example, that technology has grown, that oil has run out, that food can no longer be imported or that other countries have started to produce goods more cheaply.

WEST AFRICA: A GROWING POPULATION

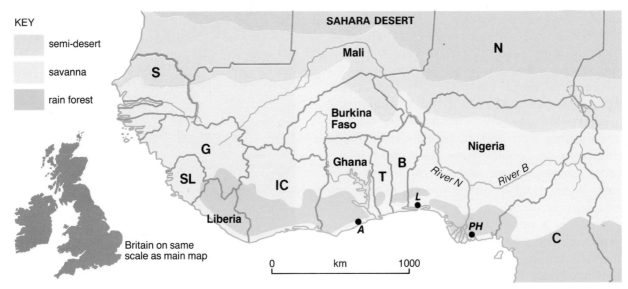

28 West Africa

In some parts of the world the population is increasing rapidly, as in West Africa, shown in map 28. The countries in these areas may not yet be as crowded as Britain or Japan but they are also not as rich. It is harder for them to raise the living standards of their people as the population grows fast. It may be difficult for you to understand and imagine what life in these countries can be like: the landscape, weather, work and lifestyle are probably very different from those that you know.

In the area shown on map 28 there are two main types of landscape, each very different. Photograph 29 shows the dry savanna grasslands which are found in the area south of the Sahara Desert. Photograph 30 shows the lush green tropical rain forest which is found in some of the coastal areas.

Here is a description of savanna landscape.

'Wide and vast; flat and rocky; bushes, thorns and twisted trees; brick-coloured bare earth; winding animal and human paths of beaten earth; insects buzzing; fierce light; shady earthen houses grouped in scattered villages; space and quiet; sudden storms and huge starry night skies; people gathering and fetching water, wood and food; animals wandering; little scraps of farms.'

29 Savanna land

Core work

44 Trace map 28 or ask your teacher for a copy of it. Use an atlas to find the names of the countries, rivers and towns initialled on the map. Write them on your map.

45 (a) Read the description of savanna landscape. Write a similar list of words and phrases which would describe the scene around your school or home.
   (b) Write a list of words which would describe the scene in photograph 30.
   (c) How are the areas in photographs 29 and 30 different from the place where you live?

Extension work

46 (a) Make a tracing of Britain from map 28 (small map).
   (b) Work out how many times it would fit into (i) Nigeria (ii) Ghana.

47 What do photographs 29 and 30 suggest to you about the climate in West Africa?

30 Tropical rain forest

Life in a savanna compound

In many parts of the savanna, families live in small farming villages or compounds (see plan 31). All the people in the compound are from the one family and may cover three or four generations. The men and women live in separate houses. The houses are round, about 3 metres across, with shiny clay walls and thatched roofs. Inside there may be a raised bed or seat and a pile of soft hides. The walls may have shelves or pegs on them and are often decorated with pictures cut out of European magazines. The houses are linked by a wall which marks the boundary of the compound, like a fort. Seats are built into the wall. For most of the day, life is lived in the open, round the cooking hearths, threshing floor and wells.

Women and children tend the crops, gather wood for the fires and thatch for the roofs. Young men herd the cattle, driving them into a fenced enclosure in the evening for safety. Men often go to the towns to work for extra money. The farms do not produce enough to pay for necessary taxes, schooling, clothes and medical care.

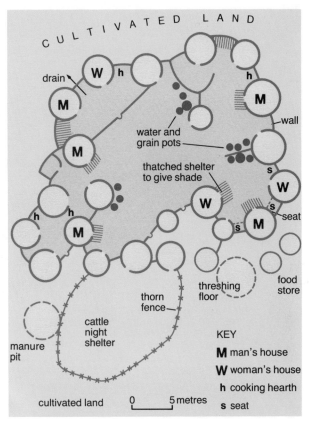

31 A plan of a savanna compound

32 A savanna village in Mali, from the air

In the dry season girls may walk 1–5 km to collect water from a dry river bed, digging holes and waiting 1–4 hours for water to seep from the gravel to fill the water pot. More and more villages are installing wells but the pumps may break down.

There is no electricity in the villages: charcoal and paraffin are the main fuels. There are battery radios in some villages, but no television or fridges.

Services such as schools, libraries, shops, hospitals and police are found only in towns. In the villages not all children can be spared from work, nor money found to send them to primary school. Very few go to secondary school. People start work at a very early age. More and more people are leaving the compounds to work in the towns. On the farms, money is short, and the farms are too small to feed everyone. The town offers better-paid work and a more attractive lifestyle.

33 Traditional trading in the village

Core work

48 (a) Look carefully at diagram 33 which shows the way village life worked in the past. Write a short description of how villagers lived in the past.
   (b) Did villagers have much contact with the outside world in the past? How do you know?

49 (a) Look carefully at plan 31 and read the text again. Use the information to draw a scene in a compound, showing the houses with people doing their work.
   (b) Make a list of the differences between traditional and modern village life.
   (c) Make a list of the things which have not changed.
   (d) In what ways is life in a savanna compound harder than your life?

Extension work

50 (a) What are the good points and bad points about compound life in the savanna?
   (b) Usually only women and children are found in the compound during the day. Why do you think this is?
   (c) Village men find it difficult to get highly paid jobs in the village. Why do you think this is?
   (d) Why do many young people want to move to the city?

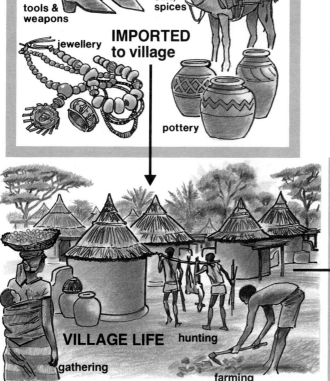

The effect of weather and landscape on farming

Farmers decide to grow certain crops, or raise certain animals, depending on the weather of the region and the type of land they have. This is true of farmers all over the world. But in industrialised countries such as Britain only a small proportion of the population rely on farming for a living. In West Africa most people grow food to eat or sell. Without it they may starve. The weather and the land are therefore especially important to them.

Study the plans of the savanna farm and the rain forest farm, and the charts which show the weather they have and the work the farmers do.

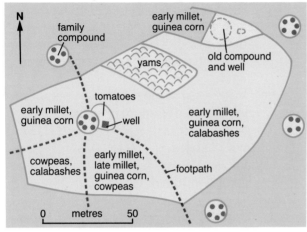

34 Savanna farm plan

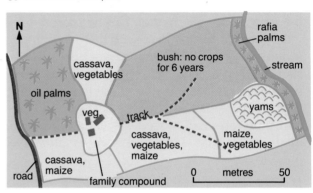

35 Rain forest farm plan

36 Savanna (north): weather and farm work

J	F	M	A	M	J	J	A	S	O	N	D
28	30	32	32	31	28	27	26	26	28	28	27
dry	dry	dry	showery	showery	showery	heavy rain	heavy rain	heavy rain	showery	dry	dry

- burn grass (J–F)
- prepare yam mounds (J–M)
- tend tomatoes (J)
- cattle drive (F–M)
- yams, early millet, vegetables (M–J)
- groundnuts and late millet (M–J)
- rice (J)
- first millet (J–A)
- main maize, millet, cowpeas, groundnuts and rice (A–O)
- cattle graze stubble (S–O)
- tobacco and tomatoes (S–N)
- tend tobacco (N–D)
- burn grass (D)

KEY to both charts: dry and sunny — showery — mainly heavy rain — general — planting — harvesting

37 Rain forest (south): weather and farm work

J	F	M	A	M	J	J	A	S	O	N	D
26	27	28	28	27	26	25	25	26	26	27	27
showery	showery	showery	showery	heavy rain	heavy rain	heavy rain	heavy rain	heavy rain	heavy rain	showery	showery

- clear bush (J–F)
- prepare yam mounds (F–M)
- yams, pumpkins, calabashes, maize (M–M)
- weed (M–J)
- first yams (A–J)
- first maize (A–J)
- beans, cassava, okra, cocoyams (J–S)
- second maize (J–A)
- general harvesting of other crops (M–S)
- main yams (O–N)
- second maize (O–N)
- clear bush (O–D)

Core work

51 (a) Make a temperature graph (like those on page 61) for the savanna farm and one for the rain forest farm, with the months along the base and the temperature on the side scales.

(b) Below the base of the graph draw rain or sun symbols like those in charts 36 and 37 for each month.

(c) Write a few sentences to describe the weather through *one year* in (i) the savanna, (ii) the rain forest.

(d) Find out what the summer and winter temperatures are for your town. Compare this with information in your graphs. Is it ever cold in the savanna or rain forest?

(e) What would be the main weather problem for farmers in savanna areas?

(f) What would be the main weather problem for farmers in rain forest areas?

52 (a) Write the headings Savanna farm and Rain forest farm. Under each heading list the products of the farm, using the information in the plans and charts.

(b) Put a tick against any crop grown on both farms. Underline any crop grown on only one of the farms.

(c) Work out from your lists which crops need a lot of rain to grow, and which crops can cope with long dry spells.

53 (a) Use the scales on plans 34 and 35 to work out how many classrooms like yours would fit into each farm. Are the farms large? Which has a large area to grow crops?

(b) Which farm has animals on it? How could they help the farmer?

(c) Which crops are grown nearest the compound (the farmer's home)? Why do you think they are grown there?

Extension work

54 At the foot of your weather graphs from Core work 51 write the following phrases where they fit the weather.
Cooler showery weather
Short heavy rains
Long heavy rains
Hot dry sunny weather
Hot showery weather

55 (a) Look at the pattern of temperature and rainfall on your graphs. Write a few sentences to describe the links you can see between cloud, rain and temperature.

(b) Which season would be the most pleasant to live in, if you could choose either farm? Why?

(c) Describe what the weather in December is like at each farm.

56 (a) Use chart 36 to work out how many months in the year savanna farmers work on (i) planting and (ii) harvesting crops.

(b) What other work do savanna farmers do?

(c) Use chart 37 to work out how many months in the year rain forest farmers work on (i) planting and (ii) harvesting.

(d) What other work do they do?

57 (a) Which jobs take longest on each of the farms?

(b) Which rain forest crops need little or no tending? (Use the plan to help you.)

(c) How is the pattern of farming – planting and harvesting – different in West Africa from farming in your area?

58 Look at the names of the crops grown on the farms. Write them down. Try to find out what they look like, how they grow and how they are used.

59 West African farmers tend to have very few machines. Find out what tools they use, and try to explain why they have fewer machines than farmers in many other countries. Write your ideas down as a report.

Population change in Nigeria

The population of Nigeria is growing fast. There are many more children and young people than there are older people.

In the countryside, larger families have meant that there is often not enough food to go round or enough to sell to buy clothes and tools, and to pay for education. In the 1970s Nigeria had oil to sell abroad (at good prices) and the country became richer. There were jobs in the oil industry and at the ports where goods were imported and exported. Living standards in the cities became much higher than in the countryside. Many people from the countryside moved to Lagos, the main city and port of the country.

38 Lagos

Core work

60 (a) Use Table 2 to draw a bar graph showing Lagos' population from 1850 to 1980. Describe the growth pattern.
   (b) Why did families move to Lagos? Look at Table 3, photograph 38 and the text to help you.
   (c) What problems might Lagos have if very many more people move there in the future?
61 (a) Why would a Nigerian farming family want to move to a city?
   (b) What might they miss about village life?

Table 2 Population of Lagos

Year	Population
1850	20 000
1900	40 000
1921	99 700
1931	126 100
1952	267 400
1963	665 250
1972	2 100 000
1980	3 600 000

Extension work

62 Look at photograph 38. Do you know any other city which looks like Lagos? Is it as you expected it to look?
63 Oil prices have fallen in the 1980s. How might this have affected jobs in Nigeria? Give reasons.

Table 3 Housing conditions in two Nigerian cities

City	Percentage of households living in one room only	Average number of persons per room	Percentage of houses with tap water	Percentage of houses with flush toilets	Percentage of houses with electricity
Lagos	72.5	3.8	71.7	43.5	93.2
Kano (an older city in northern Nigeria)	69.1	2.4	26.1	1.8	69.1

6 Aid and trade

Introduction

Not all parts of the world are equally rich. Some have harsh weather which makes farming difficult. Some have no timber, metal ores or fuels to sell to other countries or use to build up industry. Some have to cope with earthquakes, floods or drought which devastate their farms and towns. Others have too many people to feed, clothe, house and educate from their limited resources; and their populations are rising fast.

On the other hand, there are countries with successful industries and productive farms, which have many goods to sell abroad and whose populations are rising very slowly or not at all. These countries are the lucky ones.

In this chapter we shall look at the problems that poor countries face, and the different ways in which these countries can be helped, or can help themselves, to solve these problems. Richer countries can play a large part in this, but solving the problems may be more difficult and take longer than anyone might have imagined.

AID

Emergency aid

A dictionary will tell you that to 'aid' someone means 'to come to the help of' someone in difficulty. Rich countries in the world give aid to the poor countries to help them become wealthier.

There are many kinds of help a poor country may need, depending on the problems it is facing at the time. The most urgent help it needs is when a natural disaster strikes.

Let us look at an example of such help. When a natural disaster such as drought happens, crops and animals die, food runs out and people are left with no means to survive. They leave their homes in search of food and help. This has happened in the 1980s in Ethiopia. Ethiopia is a poor country and could not cope with such a large-scale disaster on its own. Mexico was in a similar position when an earthquake struck in October 1985 (see pages 16–18). Diagram 1 shows how money can be raised and aid sent to poor countries when disaster strikes. We shall look in more detail at how this affected the Ethiopian drought and famine later in this chapter.

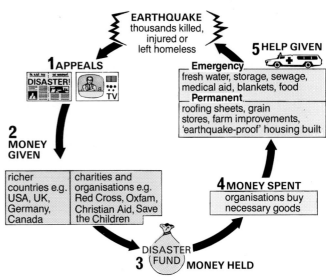

1 Raising money and sending aid for an emergency

Core work

1. What problems might people face after an earthquake? (Clues: shelter, water, food, disease, wounds.)
2. (a) Look at diagram 1. Describe the different ways in which money can be raised for a disaster, and the ways the money can be used to help the victims.
   (b) International organisations such as Oxfam often give permanent help to poor countries, as well as emergency relief. Look at the items of permanent help listed in diagram 1 and explain how each would benefit the people for the future.
   (c) Why is permanent help needed as much as emergency aid?

2 Famine areas in Africa

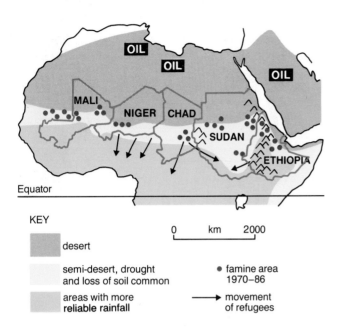

KEY

- desert
- semi-desert, drought and loss of soil common
- areas with more reliable rainfall
- • famine area 1970–86
- → movement of refugees

Long-term aid

Population in the part of Africa shown in map 2 is growing very fast, yet the harvests as a whole increase slowly, and often fail because of drought. There has not been enough money to plan for bad years, and governments have spent money on things such as fighting civil wars, instead of building roads, planting forests and improving farming or water supply. Over 20 million people in the five countries shown on map 2 are now affected by drought and famine. The problem is too serious and too long-lasting for the countries to handle on their own.

In October 1984, television pictures and reports on the famine in Ethiopia made many British people aware for the first time of how serious the problem was. At Christmas time, a large group of well-known pop singers (Band Aid) made a record to raise money. This project was so successful (£8 million was raised) that a relief organisation named Band Aid was formed. In the summer of 1985 they raised over £50 million through a huge Live Aid concert shown on television throughout the world with appeals for donations. In Britain alone, other organisations and charities raised £70 million in seven months, to be sent to the famine areas. This was used to buy food, shelter and medical supplies. In 1985 the autumn rains in the famine areas were better than for some years, and harvests improved; but the problems continue and, if they are ever to be solved properly, a different kind of aid is needed to help these countries to look after their people.

Core work

3 (a) Look at map 2. Choose *one* of the countries shown. Write a description, based on information from the map, of the different problems the country faces.

(b) New look for your chosen country in Table 1. Write down the extra information which the table gives you about the country. What problems can you guess the country faces from this information?

(c) What do governments need to spend money on to improve things?

(d) How did British people become involved? How did they help?

Extension work

4 (a) Use library books or atlases to find out as much as you can about your chosen country. Write a report of your findings.

(b) Does your report confirm your earlier information or not?

Table 1

Country	Millions of people	Life expectancy	Average wealth per person (1985)	Change in harvest 1975–85
Mali	7	45	£ 110	up 6%
Niger	6	45	£ 180	up 22%
Chad	5	43	£ 120	up 1%
Sudan	21	48	£ 300	down 6%
Ethiopia	41	43	£ 90	up 6%
UK (for comparison)	56	72	£6200	

Although Ethiopia badly needs emergency aid such as food and medical supplies, this only helps in the short term. Diagram 4 shows some of the problems that are preventing Ethiopia from getting over the severe famine and poverty that its people face. Diagram 5 shows some long-term help that could be given to tackle and solve these and the climate problems.

3 Famine in Ethiopia

4 Present problems in Ethiopia

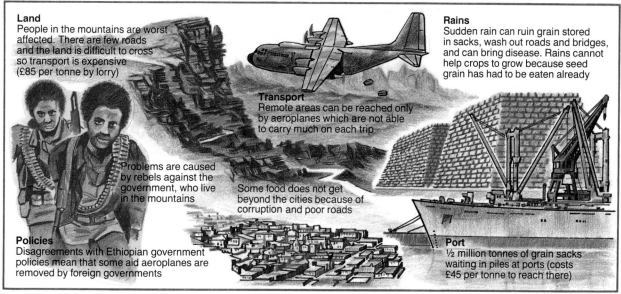

Land
People in the mountains are worst affected. There are few roads and the land is difficult to cross so transport is expensive (£85 per tonne by lorry)

Rains
Sudden rain can ruin grain stored in sacks, wash out roads and bridges, and can bring disease. Rains cannot help crops to grow because seed grain has had to be eaten already

Transport
Remote areas can be reached only by aeroplanes which are not able to carry much on each trip

Problems are caused by rebels against the government, who live in the mountains

Some food does not get beyond the cities because of corruption and poor roads

Policies
Disagreements with Ethiopian government policies mean that some aid aeroplanes are removed by foreign governments

Port
½ million tonnes of grain sacks waiting in piles at ports (costs £45 per tonne to reach there)

5 Possible long-term solutions to these problems

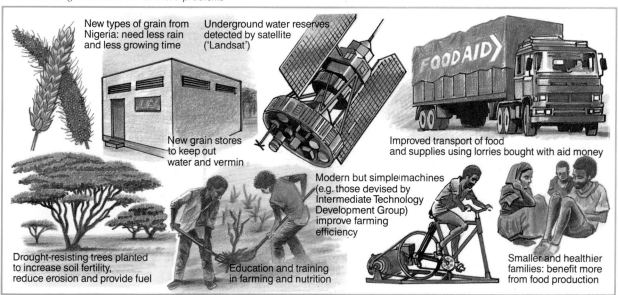

New types of grain from Nigeria: need less rain and less growing time

Underground water reserves detected by satellite ('Landsat')

New grain stores to keep out water and vermin

Improved transport of food and supplies using lorries bought with aid money

Modern but simple machines (e.g. those devised by Intermediate Technology Development Group) improve farming efficiency

Drought-resisting trees planted to increase soil fertility, reduce erosion and provide fuel

Education and training in farming and nutrition

Smaller and healthier families: benefit more from food production

Core work

5 (a) Use diagram 4 to make a list of the problems there are even after grain supplies reach Ethiopia.
  (b) Beside each problem in your list write down your suggested solution to the problem.
  (c) Read your list of solutions. Would they cost much money? Can you think of ways to save money but still solve the problems?
  (d) Now use diagram 5 to make a report on how the long-term problems of the area can be solved and the people helped to feed themselves.

Extension work

6 (a) Look back at photograph 3. How do you imagine the people in the photo felt when it was taken?
  (b) Why do people in richer countries never get as hungry or as poor as this?
7 (a) Why is long-term aid a better use of money than short-term emergency aid?
  (b) Why do you think so much aid money gets spent on emergency aid in that case?
  (c) Are there any problems you can think of which would be caused if people had to rely on aid all the time?

Where does aid come from?

Most aid in the world is given directly by one country to another. This is know as **bi-lateral aid**. Sometimes political reasons (rather than social or economic reasons) are involved when this type of aid is given, so the money is not always used in the best way. The other way aid can be given is indirectly through the World Bank, an agency of the United Nations Organisation. This is **multi-lateral aid**, and diagram 6 shows how it works. This kind of aid is not a gift. In most cases the borrowing country pays interest for the money it is given.

Depositers pay in money and earn interest (extra money)

Borrowers convince bankers that money loaned to them will be well used to make money and pay back the loan plus interest.

6 How the World Bank works

Table 2 shows the amount of aid given by some rich countries to the poor countries of the world. There are many more poor countries in the world than rich ones. The lucky richer countries need the help of people in the poor lands to provide them with food and raw materials, and to buy their goods. It makes sense for the richer countries to help these countries become wealthier so that they can afford to buy more goods.

On the other hand poor people can be desperate people. If those in the rich lands do not learn to give, then the poor may well learn to take.

Table 2 Official aid given by some richer countries to poorer countries in 1981

	Total amount given per head of population (pence)
USA	20
France	50
Germany	47
Japan	18
Saudi Arabia	200
UK	20
Netherlands	70
Kuwait	533
Canada	26

Core work

8 (a) Look at diagram 6. Which countries pay money into the World Bank?
  (b) Which countries want to borrow money from the World Bank?

(c) If you borrow money from a bank you must pay interest to the bank. What is interest?
(d) How do you think the borrowing countries hope to make money to pay back the loan and interest?
(e) Suppose their plan does not work out. What problems would they then face?

9 (a) Draw a bar graph to show the figures in the right-hand column of Table 2.
(b) How does the UK compare with other countries in the amount of aid each person gives?
(c) What could you buy with the amount of money each person in France, Canada, UK or Japan gave in 1981?
(d) How much do you think would be a fair amount for each person to give each year? Why?

TRADE

Sometimes a country is lucky and something it produces is (or becomes) very valuable. Other countries want to buy the product so the country becomes richer. This is especially true where a metal ore or fuel is discovered underground when no-one knew it was there previously. Money earned this way comes from **trade**: the buying and selling of raw materials or manufactured goods.

Trading in oil

Oil is only found in certain parts of the world. It is in great demand, especially in industrialised countries, for fuel for trains, aeroplanes, boats, cars and lorries, to produce electricity for industry and to make plastics and fibres. Many of the countries who want to use it most (see map 7) do

7 Oil producers and users: the patterns of trade

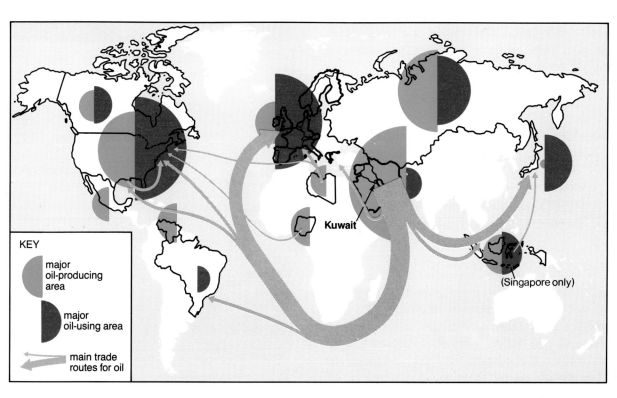

not produce it (or enough of it) themselves, so they must buy it (**import** it) from other countries. Most of the countries with good supplies of oil are (or were) poor countries. They have allowed oil companies to extract and sell (**export**) their oil abroad, making large profits for themselves and the oil companies. This is what happened in Kuwait.

Kuwait (see map 7) is a small country on the shores of the Persian Gulf, surrounded by larger and more powerful nations. Kuwait is a desert country and, before oil was discovered there in 1946, most people made a living out of herding sheep and goats from one oasis to another. Life was short and comforts few. For some Kuwaitis this is still true but, for most of the them, life is now very different. The discovery of oil has made Kuwait a rich country.

Kuwait has been fortunate to find a valuable resource like oil, but developing an oil industry is not always straightforward. It can be like a game of chance, played for high stakes. (You can try this game later, on page 106).

Kuwait met many of the problems you will face if you play the game on page 106. Only just over a million people live in Kuwait, and there were few industries there to use oil when it was discovered in 1946. Even today, when production has been increased tenfold, the oil is mainly exported because there is still little industry in Kuwait itself.

off in Kuwait, since possible markets for the gas are too far away. One use it does have, however, is to help make fresh water for the people of Kuwait.

There is little fresh water in Kuwait, so a **desalination plant** has been built to remove salt from sea water so that the water can be used in

8 Laying an oil pipeline

Core work

10 Make three lists of areas, using an atlas and map 7: one of oil-producing areas; one of oil-using areas; and one of areas which both produce and use oil.
11 (a) Which countries sell most oil?
   (b) Which are the main buyers?

Photograph 8 shows an oil pipeline being laid, and map 9 shows where the main oilfields and pipelines are in Kuwait.

Natural gas occurs with oil in Kuwait, as it does under the North Sea. But, whereas in Britain the gas is used as a fuel, most of it is burnt

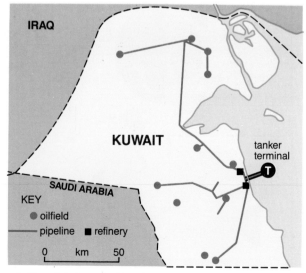

9 The oil industry in Kuwait

10 A desalination plant, Kuwait

homes and industry (see photograph 10). In the desalination plant, natural gas is used to heat sea water so that it evaporates. The steam, or water vapour, is cooled (condensed) to form pure water, and the salt is left behind as a deposit. The plant was built by a Scottish company, Weir Westgarth of Glasgow. It provides homes, offices and industries with badly needed water, but the water is costly to produce, so there is a limit to the amount which can be made. This shortage of water in turn limits the amount of industry which can be developed in Kuwait, because most industries use water.

A difficult problem faces the Kuwait government. There are more people in Kuwait than before and these people are wealthier than before. All this is thanks to oil. But little industry has developed, partly because of the small population, and also because Kuwait has no other raw materials or resources which would attract industry into the country. 'What will happen when the oil runs out?' is the vital question.

Core work

12 (a) Use photograph 8 to describe the landscape and climate in Kuwait. Which aspects of the landscape would make the building of an oil pipeline difficult, and which would make it easy?
   (b) The pipeline shown in photograph 8 is above ground. In other parts of the world pipelines are sunk below the ground. Suggest one advantage and one disadvantage of each of these methods.
   (c) Why do you think a pipeline is the most efficient way of transporting oil across Kuwait?
   (d) Where do most of the pipelines in Kuwait lead to? (See map 9.) Why do you think this is?
   (e) Not all oil is exported in its crude form from Kuwait. What clue is there for this in map 9?

Extension work

13 (a) Why is a desalination plant useful for the people of Kuwait?
   (b) Why do you think the Kuwait government were not able to build the plant themselves?
14 (a) What problem do the Kuwait government face?
   (b) Why do you think industrialised countries such as Britain prefer to buy crude oil from Kuwait rather than products made from oil (for example, plastic)?

The oil game

The aim of the game is to find and develop oil as soon as possible and for as long as possible.

Rules
1. Play in groups of up to 4.
2. Use a die and a marker or counter each.
3. Each player chooses a base.
4. Each player throws a die till he/she throws the number of his/her own base, then moves to the Strike Oil square next to the base.
5. Each player throws a die and moves round the appropriate number of squares in a clockwise direction.
6. Follow any instructions printed on a square that you land on, and note down the development (e.g. Oil find small).

Bridges carry you across to next circuit.

 Lucky squares mean you can move immediately across any one bridge you choose.

 Time square: count 10 years each time you pass it.

At the end of the game, write down
(a) how many 'years' it takes you to get to the middle circuit, and the various events or developments that happened to you on the way there;
(b) how many 'years' it takes you to get to the centre point, and what you had to do to get there.

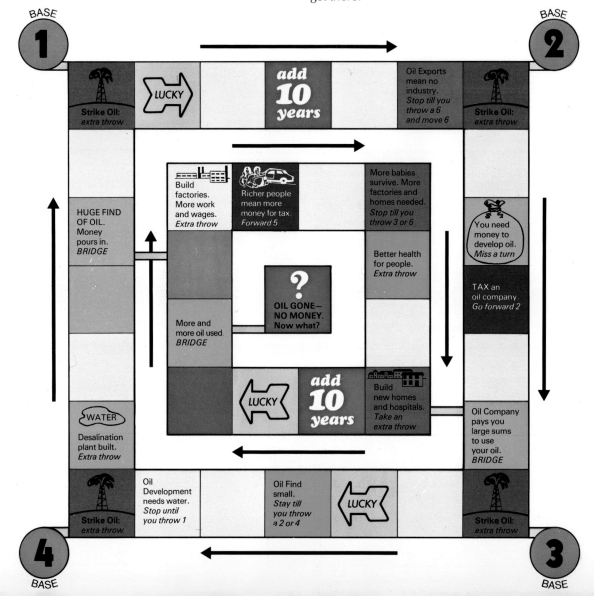

Fair trade for rich and poor

There are few wealthy industrial countries (such as the USA, Japan) in the world, and very many poor countries which rely on selling raw materials to earn their income. The raw materials bought by wealthy countries can be used to make other products which, in turn, can be sold or used to create more benefits. For example, oil is used to make fertilisers which help to increase crop yields. Poor countries, such as Ethiopia and Guatemala, where crop yields are very low, cannot afford to buy much oil or to process it to make fertilisers. Their need is much greater, but their income is much smaller.

Photographs 11 and 12 show particularly useful British exports being made and being used abroad. The factory where the tractor is being made cost millions of pounds to build and equip. Poor countries often cannot afford this investment. The tractor is useful in these countries to improve farming. If farming is improved, the farmers and the country can become wealthier, and eventually perhaps build their own tractor factory, when there are enough wealthy farmers to make it worthwhile.

11 Tractors being made in a British factory

12 A British tractor being used in a poor country

To become wealthier, a poor country can do several things:
- charge more for the materials it sells,
- sell materials as cheaply as possible to encourage other countries to buy more,
- stop selling the materials, and keep them until factories are built to make use of them,
- forget about industry and concentrate on farming to get richer from the land,
- ask for loans or grants from richer countries to pay for building the factories.

Core work

15 (a) Describe in your own words how rich countries can become even richer through trade.
   (b) Why is it difficult for poor countries to do the same thing?
   (c) Why is tractor-manufacturing good for Britain?
   (d) How can buying tractors help a poor country get richer?

Extension work

16 (a) A telex message arrived at the Department of Trade and Industry from a foreign country stating a list of some British goods it wanted to import. The letters in the words had become scrambled because the machine did not operate properly. What did the country wish to import from Britain?

   nsweapo   racs   taicrfar   alschemic
   seuocmptr   rtatrcos

   (b) Why would they wish to import these goods rather than try to make them themselves?

17 Trying to improve trade is a major problem for poor countries. Read the solutions listed in the box above, then choose two that you would follow if you were a leader of a poor country. Why would you choose these rather than the others?

World patterns of trade

Countries buy goods they need and sell those they produce to make money to improve the standard of living of their people. For example, Britain imports a lot of its food, and tea and coffee which only grow in hot climates. It also imports manufactured goods such as cars, televisions and stereos, as well as metal ores and equipment for its own industries.

Rich countries have many resources and a variety of products to sell. If trade in one of these is poor for some reason, the country is not badly affected because the other export products continue to sell well. Poor countries usually have less to trade and often rely on selling one main resource or product only. This means that if trade in this product is poor the country loses a lot of money. Yet these countries are often the ones most in need of money to solve emergency problems. Map 12 shows the worldwide variety of trade.

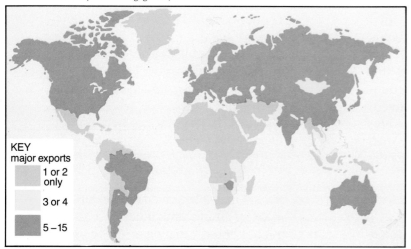

13 Variety of trading goods, 1980

Table 3 Some countries which rely on one main export

Country	Main export	Percentage of total trade
Nigeria/ Saudi Arabia/ Oman	oil	95
Cuba/ Mauritius	sugar/ honey	80
Uganda	coffee	97
Zambia	copper	87
Ghana	cocoa	74
Jamaica	bauxite	78
Mauritania	iron ore	83

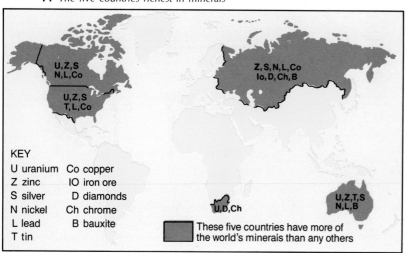

14 The five countries richest in minerals

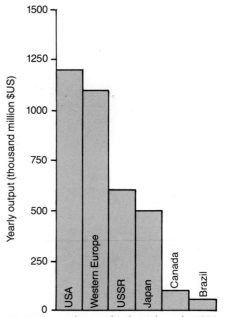

15 Major producers of industrial goods, 1970s

Core work

18 (a) Use map 13 and an atlas to find out whether each of the following countries has many kinds of goods to trade or very few.
UK, USA, Saudi Arabia, Brazil, Egypt, India, Nigeria, USSR.
(b) Look at Table 3. What would happen to the income of Uganda if coffee suddenly became unpopular and could not be sold abroad?
(c) Map 14 shows the five main mineral-producing countries. Which of the countries shown produce iron ore? Why would they not suffer as much as Mauritania (see Table 3) if the price of iron ore fell?

19 (a) Look at map 14 and an atlas. Which country produces the most kinds of minerals?
(b) What are the names of the other four main producers?

Extension work

20 Why is it dangerous to depend on one crop for export as much as Uganda and Ghana do (see Table 3)?
21 Look at graph 15 and maps 13 and 14, and say what links you can see between the patterns they show.

Rich and poor

As we have seen, the rich countries are the successful traders. The poor countries are the ones with fewer or less valuable goods to trade.

Map 16 shows the general pattern of wealth and poverty in the world (although there are of course rich and poor people in all countries). The rich industrialised countries are often called the First World; the communist of Eastern Europe and the USSR are the Second World; the poorer countries of the 'south' are called the Third World.

16 First, Second and Third World countries

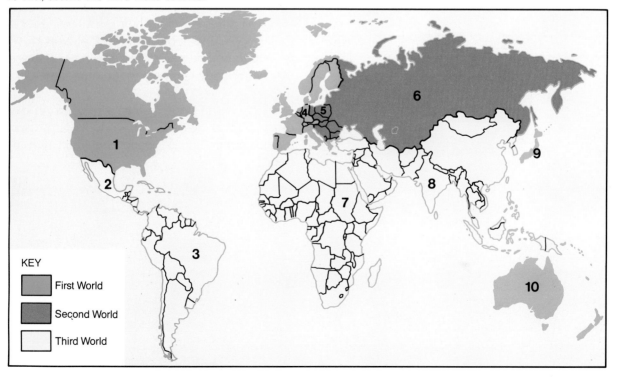

Core work

22 (a) Is a First World country more or less rich than a Third World country?
   (b) Use an atlas and map 16 to find the names of the countries numbered 1–10. Beside each one say whether it is a First, Second or Third World country.
   (c) Which continents have (i) most First World countries, (ii) most Third World countries?

Extension work

23 (a) Describe what the graphs in 17 show.
   (b) Which group of countries has (i) the biggest share of the world's population, (ii) the biggest share of the world's wealth?
   (c) What effect will this have on the standard of living of people living in (i) the First World, (ii) the Third World?

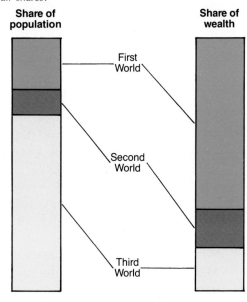

17 Fair shares?

Summary

The world's wealth is not equally shared among its people. Poor countries face difficult problems which take time, money and resources to solve. Nature often adds to the problems (or creates them), especially in areas where natural disasters are common.

In this chapter we have seen how aid can help poor countries develop their resources, educate their people and control their population growth to help them become richer. We have also seen that better trade can help them to become richer too.

The rich and poor countries depend on each other a great deal. If the world is to prosper, better cooperation is needed between rich and poor.

7 Using and misusing the environment

What is the environment?

Before you read any further, look up and look around you. What you see, both inside and outside the room, is your **environment**. Traffic, buildings, people, animals and trees are all part of the environment. Some of the things you cannot see are also part of your environment: for example, the air you breathe.

The other chapters in this book have all described environments in various parts of the world. There is a wide range, from urban environments like Hong Kong, to environments like deserts where there are few people and the landscape is practically untouched.

1 The environment

Whatever the environment, it is likely to be made up of some or all of the following elements.

- The **physical (natural) landscape**: the scenery of the land as made by nature (hills, valleys, streams, forests, wildlife).
- The **human landscape**: the scenery as changed by people (farms, settlements, industries and communications).
- The **cultural landscape**: the people who live in these surroundings (the jobs they do, the life they lead, the problems and interests they have and the decisions they make).

Diagram 1 summarises what we mean by 'environment'.

Not only do environments differ, but the ways in which we describe them can also vary. Extracts and illustrations 2–5 show four examples of ways in which different people describe environments.

From what you have seen in the earlier chapters of this book, you will have noticed that most environments in the world are constantly changing. The changes are made by people and by nature. A volcanic eruption changes the environment and affects people's lives. A caravan park changes the environment and also affects people's lives. Although there is not much we can do to prevent change by nature, we can do a great deal to control changes made by people.

Decisions made now are the decisions which will affect the lives of people in the future. **The problem is knowing which are the right decisions to make**.

3 An author's description of part of Sheffield

> The problems of Wybourn housing estate are simple: the area has nothing – few shops, no church, no secondary school, no social centre . . . True, it has a youth club but this is on the fringe of the estate and many parents will not allow their children to make the hazardous journey to it; within the last month a two-year-old girl was run down and killed virtually on the youth-club doorstep. Apart from row upon row of terraced housing the only other buildings in the area are three fish-and-chip shops and a large primary school whose amenities (such as they are) are denied the children after four o'clock.
>
> The streets are their only refuge. It is not surprising that petty pilfering is common and the general crime rate extremely high.
>
> The kids' horizons are restricted within the boundaries of the estate.

4 Part of a tourist brochure for Tamble Haven

> **Tamble Haven**
> You can't believe it's true till you visit Tamble Haven. The peace and tranquility in this forgotten corner of old England will stay with you for a lifetime. Come and pause a while beside the busy little harbour, or wander through the cobbled streets of the historic village with its red-roofed cottages and potters' workshops. Or visit the church – so steeped in memories of the past and reminders of fishermen's lives given to the sea.

5 A wildlife reserve

2 'My environment': a painting by a 10 year old

Core work

1 Make a copy of diagram 1, or use it to help you make a collage of your own environment. Using a table as below in your workbook, list all the items in three columns.

Physical landscape	Human landscape	Cultural landscape

2 Study extracts and illustrations 2–5 carefully.
  (a) What do these four impressions tell you about the *scale* of an environment? Does 'environment' mean small-scale local surroundings, or large-scale surroundings like whole countries? Is it always the same?
  (b) For each of the environments described, write down your reasons for finding it attractive or unattractive in a table like the one below.

Environment	Attractive? Yes/No	Reasons why
2		
3		
4		
5		

  (c) The person who wrote extract 4 was trying to attract tourists to Tamble Haven. Imagine you are writing a tourist brochure about your home environment and write a paragraph to describe it.
  Did you find that hard or easy to do? Why?

Extension work

3 How would you prefer to describe *your* environment: writing? poem? sketch or painting? photograph? Use one of these ways to describe your environment. Put the result in your workbook.

People and the environment

Here are some ways in which people use and live in their environments.

Places where people have little effect

In some parts of the world, people have little or no effect upon their natural surroundings. There they live in harmony with nature and only take from the land what they need for their day-to-day lives. As a result, they leave few signs of their activities on the landscape (see photograph 6).

6 An Aboriginal camp in Australia

Core work

4 (a) Describe the landscape that you see in photograph 6.
  (b) Describe the home of the Aborigine and suggest what it is made of.
  (c) How can you tell that these people are leading a very simple nomadic life?

5 Find out in what other areas of the world people have changed their environment very little.
  (a) Plot these areas on a blank map of the world.
  (b) What title could you give this map?

Extension work

6 Why do you think the Aborigine has been left alone to lead a simple life without too much outside interference?

Places where people drastically change things

Photographs 7–10 show some of the changes that can result from human interference. In these environments, people have used and often misused the physical environment to obtain the things they need.

8 Gullying

7 A slate quarry

9 The wind eroding soil on fields

10 On board a whale factory ship

Core work

7 Look at photograph 7.
 (a) In what ways has the area been spoiled?
 (b) When the slate quarry was first opened, people probably didn't think about the effects the spoil heaps would have on the appearance of the local environment. Nowadays, we are much more careful. What can be done to get rid of waste material from quarries or mines in useful ways?

8 Photograph 8 shows gullying in an area where the farmer has over-used the land.
 (a) How permanent do you think the damage to this area is? Give reasons.
 (b) What agents of erosion have removed the soil?

9 In the background of photograph 9 you can see another type of erosion taking place which has removed soil from fields used for arable farming.
 (a) What agent of erosion is involved here?
 (b) In what season of the year is this most likely to happen? Give reasons.

10 Remember that wildlife is also part of the environment. Photograph 10 shows whales on board a factory ship. In 1985, over 8000 whales were caught and killed to provide oil and other products, most of which can be obtained in other ways. Why do you think campaigns for the protection of whales have been raised?

Extension work

11 Discuss how the changes shown in photographs 7–10 can be prevented, or their effects made less unpleasant.

12 Find out about other creatures which are in need of protection. Make up a list and, beside each item, write down why the numbers of that creature are becoming smaller.

The environment is restored

Many of the problems we create by using the physical landscape can be put right, given money, time, effort, and the will to do it.

In Aberfan, a mining village in South Wales, it wasn't until tragedy and death had struck that the evils and dangers of piling up mining waste were fully appreciated. Here, the tip behind the village suddenly shifted and, like an avalanche, engulfed part of the village causing many deaths and the destruction of many buildings (see photograph 11). Most of the tip has now been removed, and the landscape restored (see photograph 12) – but too late to save the lives of over a hundred primary school children.

Much money is now spent on **conserving** features of the past. One such example is the North of England Open Air Museum at Beamish, Co. Durham, which shows many examples of how the people of North-east England lived and worked over the past hundred years.

Examples of transport, a railway station, a coal mine, a farm and a town have been brought here and are being restored to full working order to show people of today what life was really like in the past (see map 13 and photograph 14). Beamish is now a major tourist attraction in North-east England.

11 Aberfan before the disaster

12 Aberfan after it had been restored

Core work

13 Study map 13 and photograph 14.
  (a) Why is it important to have open-air working museums like the one at Beamish?
  (b) List some of the things you can see at Beamish.

13 Plan of the Beamish Open Air Museum

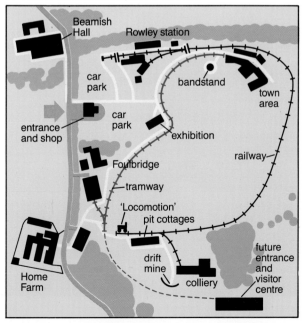

14 Inside an old shop at Beamish

A pleasant change?

In making use of nature, people can often create an environment which is attractive and gives people pleasure. Loch Faskally, at Pitlochry, is a good example. Here, a dam has been built as part of a hydro-electric scheme (see photograph 15) which provides power for hundreds of homes and factories. The dam, loch and fish ladder (see diagram 16) have all become popular tourist attractions which, in turn, have provided employment for people living in the local area.

Photographs 17 and 18 show two routeways: a motorway and a canal. These were both built to improve communications between places and to speed up the movement of goods and traffic.

Some people must have lost their land to allow the routes to be built, but these same people may also have benefited in other ways.

Making improvements to communications is bound to involve changes to the appearance of the landscape, and it is important that these changes should be attractive as well as useful. Bridges are usually large constructions and are bound to affect the appearance of an area. One example may be seen on the estuary (firth) of the River Forth in east central Scotland. This firth was first bridged by rail in 1890, then by road in 1964. Today it is possible to travel from Edinburgh to Fife either by road or rail in a few minutes and avoid the long ferry delays or lengthy detours that were once a feature of the journey (see photographs 19 and 20).

15 Loch Faskally, the dam and power station

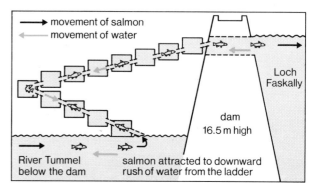

16 How a fish ladder works

18 The Oxford canal at Napton

17 The M4 motorway

19 The Forth Rail Bridge

Core work

14 Using an atlas map (or maps) of Britain, locate the following: M4 motorway, Oxford-to-Birmingham canal, Pitlochry (Scotland), Firth of Forth.
15 Study photograph 15.
  (a) Use a piece of tracing paper to draw a labelled sketch of the photograph and mark on it: the dam and power station, Loch Faskally, Pitlochry, the Perth–Inverness railway and main road, areas of woodland.
  (b) What makes this site such a good one for building a dam and a hydro-electric power station?
  (c) Suggest ways in which the loch created by the dam can be used by (i) local people, (ii) tourists (no swimming pool near the dam! Why?).
  (d) A by-pass has been built around Pitlochry recently. What benefit is this for (i) lorries travelling from Perth to Inverness, (ii) the people who live in Pitlochry, (iii) tourists who visit Pitlochry?
16 Study diagram 16.
  (a) Why is it necessary to provide a fish ladder here?
  (b) What kind of fish mainly use it?
  (c) Why should these fish want to swim upstream?
17 Look at photographs 17 and 18.
  (a) List the things that have been done to help the canal blend into the landscape.
  (b) List the things that have been done to help the motorway blend into the landscape.
  (c) Do you think the two routes have been successfully blended into the landscape?
  (d) Which of the two routes is the most attractive? Why?

20 The Forth Road Bridge

Extension work

18 Do you think the decisions to build the dam, the motorway, the canal and the bridges in photographs 15, 17, 18, 19 and 20 were the right ones? Why?
19 Write down some reasons for and against living next to a canal, a motorway, a railway, a bridge.
20 In your home area find an example of a change you like and describe it in words and pictures.
21 Find out if there are any old industrial sites near where you live. You may be able to arrange a visit to them. There are many very good ones in most parts of Britain.

Using and misusing water and air

Water and air are both important elements of the environment, and just because we cannot always *see* the effects of using them does not mean that changes are not taking place. In fact, the ways in which we use water and air must be very carefully controlled, because the effects of misuse can often be harmful, and even deadly.

Water

Photographs 21 and 22 show two different uses of the River Clyde. The sewage in photograph 21 has been specially treated to remove harmful bacteria and solids. But there are many factories which dump harmful waste and chemicals into rivers, making the water poisonous and ugly.

The paddle steamer shown in photograph 22 takes people on pleasure trips on the Clyde and on other rivers in Britain. At one time, there were dozens of similar cruises available, but very few are available today.

Air

At first sight, it is hard to imagine how aeroplanes can do anything but good. They do not clutter up valuable space on the ground in towns and cities (like railways and roads), and they help people to move long distances in a very short time. Apart from the airports they use, they don't change the appearance of the landscape.

But since the beginning of the jet age, aircraft have caused great concern. One reason is that the people who live near airports often suffer from the noise and congestion they cause. The second reason is that some groups of people are afraid that high-flying jets such as *Concorde* may seriously affect the high levels of the atmosphere and allow more harmful radiation to reach the earth from the sun.

As well as making *direct* use of the air, we also affect it *indirectly* by filling it with gases, smoke and dust (see photograph 24). This kind of **pollution** is particularly bad in industrial areas and can have many serious effects.

21 *Treated sewage being discharged into the Clyde*

22 *The* Waverley *paddle steamer in the Firth of Clyde*

23 *Living near an airport*

Core work

22 Look at photographs 21 and 22.
    (a) Why do you think treated sewage is being discharged into the River Clyde? (Think where else it might go.)

24 Industrial pollution

(b) List some of the things that might result from *untreated* sewage being put into the river.
(c) Why do you think holiday pleasure steamers like the *Waverley* have almost disappeared from our rivers and coasts today?
(d) In what other ways do people make use of rivers like the Clyde?

23 Look at photograph 23 and write down three ways in which the aircraft may have unpleasant or harmful effects on the houses and the people living in them.

24 Look at photograph 24.
  (a) What effect does this sort of industrial pollution have on the countryside and people nearby?
  (b) If the works were closed down, what might happen to the people who work there?
  (c) Do you think the smoke and fumes are worth putting up with so that people can have a job?

Extension work

25 It is often the case that changes intended to provide benefits for some may harm others. Discuss this in groups, or with the teacher and the whole class.

26 In recent years, you may have noticed reports in newspapers and on television and radio about groups of people protesting about the choice of site for a new London airport, and about the noise levels near existing airports. Find out what you can about these or similar protests in your own area. Whose side are you on? Why? *You must say why!*

27 Try to discover what laws have been passed in the last 30 years to try to cut down air pollution in our towns and cities.

28 Is your area a smokeless fuel zone? Design a poster to encourage people to stop burning raw coal in homes and factories. Who might be annoyed by your poster?

29 Find out about 'acid rain'. What is it? How is it formed? What does it do? How can it be prevented?

Ideal environments

Wherever we are, we would often prefer to be somewhere else — somewhere that we think would be ideal. Few of us are able to live in our ideal environment all the time. Some people, however, are lucky enough to spend time in their ideal environment during their leisure time in the evenings, at weekends, or on holiday.

Photographs 25 and 26 illustrate two ideal environments chosen by children. They are fortunate to have found them, for thousands of children in the world do not have any choice and may never discover their ideal. Young children may like exciting places and quiet ones. Sometimes they like to be active and noisy, and sometimes to be quiet and in an environment of their own. Do you feel the same?

Two adults chose the environments in photographs 27 and 28 as their ideals. Like children, sometimes they like quiet places and sometimes exciting places.

Core work

30 Look at photographs 25, 26, 27, 28.
   (a) What is your ideal environment? Write a description of it and say why you like it.
   (b) If you had no money, would you change your choice of an ideal place to be? If so, what would you choose instead?

25 Children enjoying the water

26 A child painting in a garden

27 People fishing

28 Crowds enjoying a sports match

Town planning

Most people live in towns and cities, so it is important that the people who build and plan them should aim to make them as ideal as possible. This is difficult. What one group would like, another may dislike. Towns and cities must provide places for people to live, shop, work, go to school and play, and also routeways to enable them to move about easily. All these things need space, but space is limited and many cities are overcrowded. In Hong Kong, for example, the problems of overcrowding are particularly severe (see photograph 29).

In many British cities, planners have tried to solve the problem of traffic congestion in the city centres by building huge new roads into and around the cities. Birmingham is a good example (see photograph 30). Here, the new roads help to keep traffic moving and make it easier for lorries and cars to reach the business centre of the city quickly and easily. But what is the price the people living in the city have to pay?

Between 1972 and 1976, the level of lead in the blood of children living near a road interchange like the one in photograph 30 more than doubled. This was caused by the lead used in petrol. If the lead level is raised much further, these children could suffer serious damage to their health, including brain damage. The government in Britain, as in many other countries, is now going to pass laws to reduce the amount of lead in petrol.

30 Birmingham from the air

29 High-rise flats and shanty homes in Hong Kong

The building of new roads in Birmingham was part of a plan to redevelop and modernise the city. Another part of this plan was to build a modern shopping centre, called the Bull Ring. This centre was opened in 1964. It is closed to traffic and consists of shopping precincts and markets. Visitors to the centre park their cars in an overhead car park and can then walk round the shops undisturbed by traffic, noise and fumes.

Pedestrian shopping precincts have been developed in most British cities and towns, and in cities in other countries. Do *you* think they are a good idea? Why?

In other cities, different solutions to traffic congestion have been tried. In Newcastle and its surrounding towns and villages a new railway system was built which goes underground in the city centre. It is called the Metro (see photograph 31), and allows people to travel to and from the city centre quickly and cheaply. It is also pollution-free, as it obtains power from overhead electric wires.

31 *The Newcastle Metro at an underground station*

Core work

31 Study photographs, 29, 30, 31.
   (a) How can you tell that Hong Kong, Birmingham and Newcastle are all short of space?
   (b) The building of high-rise flats in Hong Kong (see photograph 29) has not helped the people living in the shanty dwellings in the foreground. Why not?
   (c) What are some of the good and bad features of living in tall blocks of flats like those in photographs 29 and 30?
32 (a) How would you rather travel to school (i) by road as in photograph 30, (ii) by metro as in photograph 31? Give reasons for your answer.
   (b) Why was the railway system built underground in Newcastle city centre?
   (c) What is the Newcastle Metro railway doing to help people who live and work in Newcastle?

Extension work

33 If Birmingham's planners had known that children's health would be threatened as a result of their building new roads, what could they have done? What problems would occur if the roads had not been built?
34 (a) Does the city or large town nearest you have traffic congestion?
   (b) If 'yes', how are planners trying to improve the flow of traffic?
   (c) If 'no', how have the planners solved the problems?
35 Write down or draw plans of how you would solve traffic problems in a large city. Would you do any of the following: ban cars in the city centre; build multi-storey car parks; build motorways to the city centre; build underground railways; use only buses or trams; have free parking or no parking?
36 Many cities are now demolishing multi-storey blocks of flats that were built only 20 years ago. Find out why.

32 A gable-end painting in Glasgow

The street environment

In an area of old housing in Glasgow, where houses have been torn down to make room for roads, some of the gable ends have been painted to brighten up the area (see photograph 32). The Scottish Arts Council, in 1974, asked Glasgow artists to send in designs for gable end paintings, and the best ones were then used. Do you think they improve the environment?

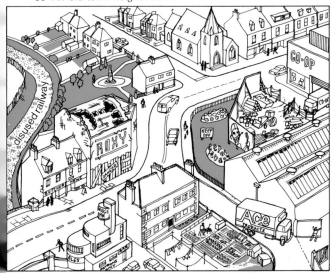

33 An old town neighbourhood

Core work

37 Imagine you live in the neighbourhood shown in diagram 33.
   (a) What is good and bad about this area?
   (b) What parts of it could be improved?
   (c) How could they be improved?
   (d) Who could be asked to make the improvements?

Extension work

38 Do a painting designed to cover any large blank surface or gable end in your street, or any street. A group of you could send your designs to the local council, newspaper or Arts Council.

A rural environment under threat

For many city dwellers, their ideal environment for a holiday or a place to retire, is a country area of unspoilt natural beauty within easy reach of the city. Lake Tahoe, near San Francisco in California, is such a place (see photograph 34 and map 35). Thousands of tourists visit Lake Tahoe to enjoy the scenery and use the lake for boating, etc. Why is the area under threat? Here are some of the reasons. Think about the effects they would have.

- Many people would like to use the lake for water skiing, power boating, yachting and fishing.

34 Lake Tahoe

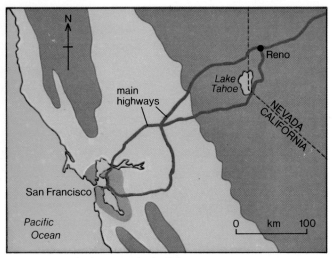

35 The location of Lake Tahoe

- Many would like to build a home on the shores of the lake, either for use in the holidays or to live in when they retire.
- Many would like a new 'fun city' like nearby Reno to be built on the lake shore. Part of the lake lies within the State of Nevada where gambling laws are much less strict than in other states in the USA. People want to take advantage of this, as has been done in Las Vegas, a gambling city in Nevada.

It is always very difficult to find fair solutions to environmental problems. There is nearly always a conflict between **developers** (who want to make money out of using an area) and **conservationists** (who want to keep an area as it is), and the result of the conflict is often a compromise, where each side has to give up something of what they wanted.

The important thing to remember is that *you*

36 People protest

have a responsibility and a part to play in shaping the environment. It is the right of every citizen to object if their environment is threatened by thoughtless actions. The group of people shown in photograph 36 are making one form of objection.

Table 1 Population increase in the Lake Tahoe area

	Permanent population	Summer peak
1950	1 600	–
1955	2 500	–
1960	5 500	30 800
1965	7 500	44 500
1970	9 500	58 500
1975	15 000	75 000

Core work

39 (a) Look at an atlas map of the USA. Find Lake Tahoe, San Francisco, Reno and Las Vegas.
   (b) Why would it be difficult for the lake to be used freely in all the ways mentioned?
   (c) Using the figures in Table 1 draw a graph, using one colour of line to show the permanent population and another colour of line to show the number of summer visitors to Lake Tahoe.

40 Study photograph 36.
   (a) What is the word used to describe this kind of objection?
   (b) What other ways could these people use, and to whom should they complain?

Extension work

41 Imagine that you are *either* a conservationist (someone who wants to keep an area in its present state), or a developer (someone who wants to make money out of using an area). Write down your reasons for wanting to conserve or develop the Lake Tahoe area.

Instead of writing your case, you could make it a topic for debate in your class.

Plan your future

Here is your chance to make some decisions. Whatever you decide to do, remember that changes to the environment should only be made if they are for the benefit of *people* and the *surroundings* in which they live.

Picture 37 shows part of a landscape containing:
- a large city with port and airport, Inverbeath;
- a small fishing port and holiday resort, Abermoss;
- a large area of good farming land, broken only by a range of low hills, the Penthills;
- an area of mountains in the distance, the Snowgorms.

Imagine you live in the small village of Abermoss (population 3000). Most of the people who live here either
- work locally in fishing, or the industries serving the village: shops, offices, garages, hairdressing, etc.,
- or are retired, and have moved to Abermoss because of the attractive countryside of the area and the view out to sea,
- or travel to work every day to Inverbeath.

You learn that an international oil company – the Organisation for Petroleum, Oil and Natural Gas (OPONG) – is intending to build an oil refinery on the coast next to Abermoss, with a large tanker terminal (for exporting oil) offshore. This is because the coast is close to offshore oil wells which have recently come 'on stream'. Many of the residents of Abermoss are alarmed by the proposals, and quickly form an association to fight these plans. Their fears are as follows:
- The scenic beauty of the area will be destroyed by this industrial development.
- There will be air pollution from the oil refinery: no more fresh sea air.
- At any time, there could be an oil spillage, ruining the cliffs, beaches, and coastal wildlife.

37 The Inverbeath–Abermoss area

- There will always be a risk of explosion or fire at the refinery, which could damage houses in Abermoss, and even cause injury or loss of life.
- Abermoss will change from a pleasant coastal village to a large, noisy industrial town, as more and more industries are attracted to the area.

Core work

42 Study picture 37 carefully.
   (a) In what ways do the people of Inverbeath use the surrounding countryside?
   (b) In what ways might the farmers and their families use the city?

43 (a) What environmental problems might Inverbeath have? (Hints: roads, airport, power station, port, offshore oil platforms.)
   (b) How can Inverbeath solve these problems?
   (c) Would you like to see any changes take place in and around Inverbeath? If so, what? Developments? Improvements?

44 What problems do farmers living close to Inverbeath face
   (a) now? (Hints: traffic, day-trippers, dogs, country code.)
   (b) in the future? (Hints: housing, industry, motorways.)

45 How can the farmers' problems be solved?

46 Imagine a public meeting is being held in Abermoss to discuss the planned oil refinery. The meeting will need a chairperson to control the discussion; two representatives from OPONG to outline the plans, and answer questions; and residents of Abermoss. The chairperson may have difficulty keeping the meeting under control at times, as many of the residents will be angry, and some will be anxious. The oil company representatives will try to allay some of the fears of the residents, but from time to time they may receive shouts, boos, etc., so that it will be difficult to hear what they are saying. Some of the residents will welcome the refinery, as it will provide more business and more jobs, but others will be very worried. The people in Abermoss will come from all walks of life, e.g. shopkeepers, hotel/guest-house owners, pub owners, retired people, fishermen, ministers, unemployed school leavers, housewives, doctors, teachers, etc.

Now hold a public meeting in class with everyone playing different parts. You will need to have roles to play and agree to do some homework preparing your parts, deciding what you will do and say.

Notes for the teacher

Rationale

The Outlook Geography series, (*Our Landscapes* and *Worldscapes*) was designed to provide a rounded course in geographical studies for 12–14 year old pupils, a course which might provide such pupils with their only contact with the skills and ideas of the subject and the insights it can give into contemporary phenomena and problems.

Taken together, these books present an eclectic view of the contribution of geography to education. The subjects and themes covered are seen as contributing vital elements to courses in environmental studies, environmental education, social studies and contemporary studies. The narrower role of the subject as a quantitative spatial analysis is developed in ways we consider appropriate to the age, stage and wider interests of the pupils.

Our proposition is that for much of our school population, the understanding of basic ideas and straightforward relationships, and the application of a restricted range of skills in verbal language, graphics and number will be the sum of cognitive attainment. Despite such limitations, the role of people in the environment, the role of nature in the environment and the interaction of both, need to be examined and as far as possible understood.

Objectives

The writing team defines its pupil objectives as follows.

(a) *The acquisition of knowledge*
Without knowledge of context, there can be no such thing as an accurate mental map. Without the essential building blocks for thought, ideas/concepts cannot be meaningfully developed.

(b) *The development of concepts*
These include low-level concepts, developed when classes of objects or phenomena are associated accurately with the words which describe them, and such words are then used meaningfully. Some examples are volcanoes, harbours, reservoirs, irrigated farms.

They also include 'key' concepts, seen to underpin the entire curriculum and not just geography. Causality (cause and effect), change and continuity, interdependence, similarity and difference, and empathy are examples.

The geographical concepts included are those widely recognised as valid in any study of the subject. They are, for example,

1. **location:** through the use of maps in the text, and the need to use atlases; through the consideration of advantages and disadvantages of location when appropriate;
2. **scale:** not only mathematical measurement, but the microscale of specific local examples to the broadest generalisations of global patterns;
3. **relationships:** the emphasis being on the people/land relationship through correlation and causation studies;
4. **movement:** crustal movement and creative movements of water and ice; migration (push and pull); trade and the movement of goods;
5. **energy:** origins and uses of natural forms; needs for, and inequalities of, supply;
6. **change through time:** (sequent occupancy) short and long time scales of change; landscapes as palimpsests; control of change; natural and human influences;
7. **patterns and distribution:** demographic frequencies, descriptive terminology (e.g. rectangular) when appropriate;
8. **systems:** input–output in hydro-electric power production, in oil development, in valley formation.

(c) *Skills*
Pupils have to acquire and apply a number of skills. Many of these should have begun in the primary school. Pupils will need the skills to acquire information by listening, by reading to find the answers to specific questions, by looking around them at pictures, charts and graphs. They will

need to learn how to use an atlas and other books of reference, by using indexes, tables of contents, keys and locational reference systems. They will need to develop skills in taking notes, selecting appropriate facts, making appropriate classifications, and writing summaries. The skills of communicating orally are appropriate, as are skills in selecting and organising material for a chart or picture layout. Skills in translating from numbers to graphic forms and in translating from graphic form are also essential, involving as they do all the skills in map reading, photograph reading and the interpretation of graphs. The teaching of such skills, and any others as they are required, is seen as the special responsibility of the class teacher.

Teachers will note the relatively strong emphasis placed on language work; on reading for information, reading for understanding, writing descriptions and explanations. We have also indicated the need for talk, for discussions and the interchange of points of view.

(d) *Attitudes and Values*

It is doubtful whether 'correct' attitudes and values can ever be acceptably defined and agreed, far less taught. However, it may be asserted that young people must be helped by discussion, by precept and by example to form and reinforce attitudes which are helpful to and considerate of other people, and values which lead to an insistence on high quality in all aspects of life. Young people continually make value judgements; perhaps the best a teacher can do is to show how such judgements should develop from and depend on evidence; how decisions can be arrived at by balancing conflicting evidence; to discuss the reasons for, and the results of, conflicts and to show how the ability to cooperate is a most valuable attribute in a group dealing with social and environmental problems.

In *Worldscapes* we have provided opportunities for sympathetic discussion of a wide range of issues: for example, earthquakes and aid; trade and independence versus aid and dependence; the pros and cons of mining and building developments; the pressures of highly organised 'Western society' on less developed societies.

Our view of the role of these books

A book like *Worldscapes* cannot be seen as a teacher's solution to the problems of curriculum planning, resource selection and lesson design. All it can do is provide a core resource. Departments and individual teachers require to select or create other appropriate case studies to illustrate and develop the ideas, and extend the contexts. If a teacher feels that the Communist world has not been given 'fair' space, then he/she will perhaps replace the case study of North America with a case study of the development of Imperial Russia and the USSR.

As these sources are intended to be widely applied, they cannot capitalise on the value of the local environment for establishing ideas and standards for comparisons. A local river can be used as a case study in the section on river erosion. Local graveyards might exemplify the idea of change. The sizes of foreign homes may be related to the size of the classroom. Trade and aid patterns in the local community might be used to help clarify ideas of trade and aid.

It is up to the individual teacher to exploit the potentials of this book, their local environment and other appropriate resources, and integrate all of them into the planned curriculum for 12–14 year old pupils.

T.H. Masterton